Managing Burnout in the Workplace

Chandos
Information Professional Series

Series Editor: Ruth Rikowski
(Email: Rikowskigr@aol.com)

Chandos' new series of books is aimed at the busy information professional. They have been specially commissioned to provide the reader with an authoritative view of current thinking. They are designed to provide easy-to-read and (most importantly) practical coverage of topics that are of interest to librarians and other information professionals. If you would like a full listing of current and forthcoming titles, please visit our website, *www.chandospublishing.com*, email *wp@woodheadpublishing.com* or telephone +44 (0) 1223 499140.

New authors: we are always pleased to receive ideas for new titles; if you would like to write a book for Chandos, please contact Dr Glyn Jones on *gjones@chandospublishing.com* or telephone +44 (0) 1993 848726.

Bulk orders: some organisations buy a number of copies of our books. If you are interested in doing this, we would be pleased to discuss a discount. Please email *wp@woodheadpublishing.com* or telephone +44 (0) 1223 499140.

Managing Burnout in the Workplace

A guide for information professionals

NANCY McCORMACK AND
CATHERINE COTTER

CP
CHANDOS
PUBLISHING

Oxford Cambridge New Delhi

Chandos Publishing
Hexagon House
Avenue 4
Station Lane
Witney
Oxford OX28 4BN
UK
Tel: +44(0) 1993 848726
Email: *info@chandospublishing.com*
www.chandospublishing.com
www.chandospublishingonline.com

Chandos Publishing is an imprint of Woodhead Publishing Limited

Woodhead Publishing Limited
80 High Street
Sawston
Cambridge CB22 3HJ
UK
Tel: +44(0) 1223 499140
Fax: +44(0) 1223 832819
www.woodheadpublishing.com

First published in 2013

ISBN: 978-1-84334-734-7 (print)
ISBN: 978-1-78063-400-5 (online)
Chandos Information Professional Series ISSN: 2052-210X (print)
and ISSN: 2052-2118 (online)

Library of Congress Control Number: 2013941269

British Library Cataloguing-in-Publication Data.
A catalogue record for this book is available from the British Library.

Typeset by RefineCatch Limited, Bungay, Suffolk

Contents

About the authors

Nancy McCormack, BA, MA, MLIS, JD, LLM is an Associate Professor of Law at Queen's University with a background in Labour/Employment Law and Human Resources, and is former Head of the Law Library. She teaches legal research and writing to upper year and graduate law students and has published widely on librarianship and legal research. She is the co-author of *The Practical Guide to Canadian Legal Research* (Carswell, 2010) and *Updating Statutes and Regulations for all Canadian Jurisdictions* (Carswell, 2012).

Catherine Cotter, BA, LLB, MLIS, is the Reference and Instruction Librarian at the University of New Brunswick Law Library, and is a member of the Law Society of Manitoba. She teaches legal research to all first year students in the Faculty of Law and offers an advanced legal research course to upper year law students. Catherine has co-authored the *Canadian Law Dictionary* (Barron's, 2009) and *The Practical Guide to Canadian Legal Research* (Carswell, 2010).

1

What is burnout?

Abstract: This chapter discusses recent changes in the workplace which have resulted in greater sources of stress for workers. Burnout is defined and distinguished from low morale, depression and stress. A consideration follows on how burnout manifests itself in terms of changes in behavior, feelings, thinking and physical and mental health.

Key words: burnout, low morale, depression, stress, *Diagnostic and Statistical Manual of Mental Disorders*, Fourth Edition (DSM-IV), *International Statistical Classification of Diseases and Related Health Problems*, 10th Revision (ICD-10), changes in feelings, changes in health, changes in thinking, changes in behavior.

Introduction

Over the last several decades, profound changes have occurred in much of the workplace. Budgets have been cut, workers have been downsized and often fewer employees have been left to carry out the same amount of work.

As a result, the pace of work seems far more relentless than it did only a few brief decades ago. Workers are more harried than ever, yet attempting to control or limit the volume of work assigned is often futile. An individual who says no to more work might, at best, be regarded as someone who is not a team player. At worst, saying no is viewed by the organization as unacceptable, and a step towards dismissal.

Given this demanding pace, time away from work should be the period during which an employee can relax and recover from the stresses of the day. Instead, setting clear boundaries between work time and personal time has become, for many, impossible. Employees are frequently expected by their employers to check their email at work and at home, and to be reachable at any time of the day or night, including vacation. Finding a work–life balance is made more difficult as a consequence of this "constant connectivity" and results in even greater stress on the individual employee (Study, 2012).

One might think that upper management would find some way to reward employees for accepting the grinding pace and the greater intrusions of the job into their private time. Instead, in an era of shrinking resources, individuals higher up the food chain are often as worried about their own survival as the people who report to them. As a result, supervisors may take credit for projects done well but may be slow to show support when things are going badly through no fault of the workers (Maslach and Leiter, 1997). Without such support from superiors, workers find it even harder to shoulder the stresses of the job. Eventually, the volume and pace of work, along with a myriad of other factors in the workplace, break down many individuals so that they can no longer function. Burnout is often the end result.

Burned out workers are those who find themselves suffering from severe emotional fatigue, which is frequently accompanied by physiological symptoms. They feel distressed, alienated, inadequate, and unmotivated. These employees often become unrecognizable to their colleagues, particularly as they withdraw from social and other interactions. Their behaviour also changes; they may exhibit signs of depression or anger, and may turn to drugs or alcohol or employ other dysfunctional coping behaviors in an effort to deal with the stress.

Overwork, of course, is a major contributing factor to burnout. An excessive workload and an unremitting pace are no longer facts of life only in third world sweatshops. Today, even in wealthy, industrialized countries, overwork has been found to contribute to serious illness and, somewhat surprisingly, death. In countries such as Japan, for example, death from overwork even has a name – *karoshi*. Since the 1980s, 30,000 Japanese have been recognized officially as having been victims of *karoshi*, i.e., their deaths have been recorded as having come about as a result of overwork (Pannozzo and Landon, 2005).

As for other industrialized countries, while accurate information on death from overwork is harder to find, it has become more common to find statistics on workplace stress and the role it plays in serious health problems. The president of the American Institute of Stress (AIS), Paul J. Rosch, MD, FACP, says that chronic workplace stress often leads to significant health problems such as high blood pressure, cardiovascular disease, heart attacks and strokes and can aggravate many already existing conditions (Walter, 2012).

Stress and burnout also make the likelihood of workplace accidents greater. Burnout has been linked to the risk of developing back problems and other musculoskeletal disorders in the upper body (Kalia, 2002). It is known to have a connection with diabetes, suppression of the immune system, memory loss, and the shrinking of neurons in the brain (Contenta, 2010). Researchers have, in addition, found a correlation between job stress and burnout and "various self-reported indices of personal distress, including physical exhaustion, insomnia, increased use of alcohol and drugs, and marital and family problems" (Maslach and Jackson, 1981).

What does all this mean for the workplace? As these and other studies indicate, employees pay a high cost as a result of burnout. But they are not the only ones. Worldwide, the syndrome presents a serious problem for all kinds of businesses, organizations and the economy as a whole. For example, in 2002, the European Agency for Safety and Health at Work estimated that work-related stress disorders in Europe resulted in a yearly economic cost of approximately 20 billion Euros (about 25 billion US dollars) (Awa et al., 2010).

In 2005, it was reported in the press that one in five workers in Scotland thought their jobs were "highly stressful" (Gray, 2005). Several years later, in 2011, the UK's Health and Safety Executive (an independent body which conducts research and keeps an eye on work-related health, safety and illness issues) estimated that stress was responsible for 10.8 million lost workdays that year (Health and Safety Executive, 2011).

In the United States, in 2009, a survey conducted by the American Psychological Association found that 69 percent of all employees felt that their job was a significant source of stress. Forty-one percent felt stressed during a typical work day and more than half of those surveyed thought that stress had a negative impact on their productivity at work (Levinson, 2012). That same year, the US National Institute for Occupational Safety and Health (NIOSH) reported that 26 percent of workers are "often or very often burned out or stressed by their work," and 29 percent said they felt "quite a bit or extremely stressed at work" (NIOSH, 1999, p. 4).

A price tag in the neighbourhood of $300 billion per year is what the American Institute of Stress has estimated stress and burnout on the job costs US businesses (Walter, 2012). It has also estimated that more than 75 percent of all visits to doctors are stress-related (Kalia, 2002). Costs are incurred as a result of reduced productivity and revenue, decreased job satisfaction, increased absenteeism and sick leaves, job turnover, low morale, and the necessity for replacement workers, along with compensation, litigation and disability claims. Employees who are burned out are also less inclined to assist their colleagues on the job and less inclined to care about the organization or its goals. It is worth noting that next door to the US, Health Canada has also reported problems related to stress: it is responsible, for example, for an annual price tag of C$3.5 billion as a result of absenteeism by Canadian workers (Contenta, 2010).

Several continents away, a 2010 University of Melbourne and Victorian Health report found that, in Australia, job stress and depression resulting from excessive pressure in the workplace cost the economy approximately A$730 million a year. Researchers arrived at this figure by estimating the costs of lost productivity, of finding replacement workers and of medical treatment of work-related mental health problems ("Huge cost," 2010). As these figures show, burnout is a serious problem internationally.

To what extent the numbers are rising from one country to the next is not as clear. Despite the approximately 6,000 books, chapters, dissertations, and journal articles which have been published on burnout in countries around the globe over the last 35 years (Schaufeli et al., 2009), comprehensive country-wide data on burnout is not always easy to find, in part because it is not officially recognized in all countries as a specific health problem in and of itself. One country which does keep statistics – the Netherlands – has discovered that at any given time, 10 percent of employees are burned out (Senior, 2006).

Individual studies of various professional groups have also been conducted over the years and have estimated that, for example, approximately one out of every five doctors in the US and Germany is burned out at any given time and more than one in four are burned out in Great Britain (Awa et al., 2010). Reports also indicate that between 30 and 40 percent of teachers are burned out at any one time (Awa et al., 2010).

Clearly, the literature indicates that burnout results in tremendous costs to individuals, organizations and the economy. Nonetheless, the factors which contribute to burnout seem to continue unabated in the workplace, and recent surveys point to a rise in the number of burned out employees ("Jobs Mail," 2005).

While relatively few statistics exist for individuals in the information professions, the reality of the problem of burnout pertains equally to them. Not surprisingly, economic costs in these workplaces are already known to be high as a result of

> reduced employee productivity, increased sick leave, higher insurance premiums, early retirement payments, the need to recruit and orient replacement or substitute employees, provision of burnout treatment and prevention, employee sabotage, and litigation. (Nauratil, 1989, p. 4)

In this book, although the research discussed is drawn from the great wealth of literature on burnout from around the world and deals with a number of professions, we will also give special attention to the handful of studies devoted specifically to burnout in the information professions and what they say about workplace stress. The book will consider what exactly burnout is, the factors which are said to increase the risk of burnout, and how to keep burnout at bay.

To begin, therefore, the opening chapter will look at the various definitions of burnout and how experts differ on their views of the syndrome. It will review the differences between burnout, low morale, stress, and depression. The chapter will also examine how an individual feels, behaves, and thinks when burnout is on the horizon or when it has already closed in, as well as what burnout does to overall health. First of all, we turn to the problem of definition.

What is burnout?

Definition

The concept of burnout first figures prominently in print in the Graham Greene novel, *A Burnt-Out Case*. In the novel, Querry, an architect, has travelled to Africa in order to escape his former life. Although renowned in his field, Querry has lost interest in his work and in his private life. The reader comes to understand that Querry now finds his former accomplishments and relationships meaningless; he feels empty and numbed and seeks only to withdraw. He explains to a physician he meets why he came to Africa:

> . . . I wanted to be in an empty place, where no new building or women would remind me that there was a time when I was alive, with a vocation and a capacity to love – if it was love. The palsied suffer, their nerves feel, but I am one of the mutilated, doctor. (Greene, 1977, p. 46)

In 1969, Harold B. Bradley, a correctional administrator, used the term burnout to describe staff in correctional programs (Bradley, 1969). But it was not until five years later, when psychiatrist Herbert Freudenberger (1974) used burnout to describe what he saw happening to volunteers at a drug addiction clinic, that the term drew more attention. Freudenberger described how the volunteers, when starting out at the clinic, approached their work with high levels of enthusiasm and commitment. They had elevated expectations regarding what they might be able to achieve and they worked hard to realize those expectations. Within a year, however, that enthusiasm and motivation had clearly diminished, and volunteers found themselves in a state of job distress which manifested itself both mentally and physically. As Freudenberger noted, "[w]henever the expectation level is dramatically opposed to reality and the person persists in trying to reach that expectation, trouble is on the way" (1980, p. 13).

The specific trouble referred to by Freuedenberger was burnout, a term that had been used casually by workers and volunteers in the field to describe the long term effects of drug abuse on chronic drug users. Freudenberger himself now used burnout to describe the exhaustion and lack of motivation that he saw present among the volunteers after they had been working for some time. These people were experiencing a "state of fatigue or frustration brought about by devotion to a cause, way of life, or relationship that failed to produce the expected reward" (Freudenberger, 1980, p. 13). Striving beyond the mental and physical limits of their bodies, the volunteers had found themselves exhausted, depressed, and disillusioned.

Today, the term is widely used even in casual conversation to describe fatigue and exhaustion. But it is also used by hosts of researchers who study the phenomenon. In scholarly studies and empirical research involving burnout, one of the definitions most commonly used came about as the result of the work done in this area by two Americans: professor of psychology Christina Maslach and professor of human resource management Susan E. Jackson. Their definition was not, in the beginning, the product of a theory empirically tested in the field; instead, it was developed from the ground up as the result of extensive research involving various employee groups – day care workers, mental health staff, attorneys, police officers, physicians, educators, nurses, etc. – all of whom were asked to describe their jobs and work experiences.

From these studies came the definition of burnout as "a syndrome of emotional exhaustion, depersonalization, and reduced personal accomplishment that can occur among individuals who work with people in some capacity" (Maslach et al., 1996, p. 4). Burnout was seen not as the product of an occasional crisis in the workplace, but rather as the "response to chronic, everyday stress" and the "never-ending onslaught of emotional tension" which wore individuals down (Maslach, 1982, p. 11).

This wearing down of individuals and the resulting *emotional exhaustion* is the first dimension of the syndrome and is key to understanding it. As Maslach and Jackson describe it, burnout is

> a syndrome of emotional exhaustion ... that occurs frequently among individuals who do "people-work" of some kind. ... As their emotional resources are depleted, workers feel they are no longer able to give of themselves at a psychological level. (Maslach and Jackson, 1981)

Emotional exhaustion is often accompanied by a general weariness, the inability to sleep properly at night, physical lethargy, and a host of physiological symptoms including stomach problems, digestive problems, muscle fatigue, and headaches.

In burned out individuals, exhaustion is generally accompanied by feelings of *depersonalization and/or cynicism*, the second major dimension of the syndrome (Maslach and Jackson, 1981). Employees develop negative attitudes and feelings towards the recipients of their services (i.e., patrons, clients, patients, students, etc.). They also become disenchanted with their work. Communication becomes a strain, and burned out employees respond by seeking to avoid contact with the people around them.

In the library setting, for example,

> Burnt-out librarians who stay on the job ... withdraw from it in other ways – extending breaks and lunch hours, missing meetings, going home early, and routinely taking advantage of their maximum sick leave. In a library where burnout is prevalent, professional collegiality may be seriously eroded. Intra-organizational communication may break down at the informal level between coworkers, between departments, and between management and staff. (Nauratil, 1989, pp. 3–4)

Enthusiasm for the job requires energy. Where that energy is depleted, enthusiasm is replaced by cynicism, not only towards the group of people the employee is intended to help, but often towards colleagues and supervisors as well.

According to researchers, changes in personality or attitude are observable in people who are burning out: formerly even-tempered employees are suddenly short-tempered, easily irritated, rude and insensitive. They no longer feel the same commitment to the job and may fluctuate between feelings of anger, disillusionment and not caring at all. Indeed, one of the hallmarks of burnout is that individuals at risk do not notice the signs or symptoms of burnout for some length of time. One wake-up call is when they understand they are behaving in ways they do not recognize as being consistent with their own personality (Casserley and Megginson, 2009).

Finally, the third dimension of burnout is what Maslach and Jackson call *reduced personal accomplishment or ineffectiveness*. This involves a "tendency to evaluate oneself negatively, particularly with regard to one's work with clients. Workers feel unhappy about themselves and dissatisfied with their accomplishments on the job" (Maslach and Jackson, 1981). Their accomplishments seem trivial to them. They suffer from a lack of confidence; they feel less effective on the job and have doubts about their self-worth. Ironically, the more such an employee thinks about failure, the more failure begins to look like a self-fulfilling prophesy.

There have been debates about the extent to which exhaustion, depersonalization/cynicism and reduced personal accomplishment/ineffectiveness make up the burnout syndrome. Maslach notes that early studies of burnout were dismissed by other researchers as nothing more than pop psychology. But there seems little debate anymore that burnout exists and that it is pervasive. Over the last 30 years, research from around the world has contributed to what is now a large body of empirical evidence (Schaufeli et al., 2009) in addition to various theoretical models (Maslach, 2003).

The concept of burnout as a workplace phenomenon has also extended beyond individuals in the human services professions to include those who do not deal with people to any large extent in their jobs. Instead, these individuals may problem solve, mentor or work creatively in their jobs. In such cases, burnout has been defined as ". . . a state of exhaustion in which one is cynical about the value of one's occupation and doubtful of one's capacity to perform" (Maslach et al., 1996, p. 20).

Other definitions

As noted above, though the definition of burnout as set out by Maslach and Jackson is the one cited most widely in the literature, there has been considerable debate about the three components of burnout they posit.

Shirom (1989), for example, offers a definition that focuses specifically on exhaustion: "[b]urnout refers to a combination of physical fatigue, emotional exhaustion and cognitive weariness" (pp. 33–4). He and other researchers (Kristensen et al., 2005) do not view depersonalization as one of the hallmarks of burnout but rather see it as a coping behaviour which precedes burnout. Similarly, they view the inability to take satisfaction in one's job (i.e., reduced personal accomplishment) as a consequence of burnout rather than a defining characteristic.

Other researchers (e.g., Pines and Aronson, 1988) concur with Shirom that the various types of exhaustion (physical, mental and emotional) are the key determinates of burnout. Pines and Aronson define the syndrome as "[a] state of physical and emotional exhaustion caused by long-term involvement in situations that are emotionally demanding' (Pines and Aronson, 1988, p. 9). Similarly, Schaufeli and Greenglass see burnout as "a state of physical, emotional and mental exhaustion that results from long-term involvement in work situations that are emotionally demanding" (Schaufeli and Greenglass, 2001, p. 501).

As with the debate over the three dimensions of burnout posited by Maslach and Jackson, there has also been discussion over how burnout is to be measured outside of human-services jobs or professions. Researchers have indeed concluded that burnout is a problem in all areas of employment, not just the human-services area (Beemsterboer and Baum, 1984), but the instruments which measure these disparate groups are necessarily different and have produced unexpected results. For example, through the use of an instrument designed to measure the problem among those who don't necessarily work with the public, psychology professor Evangelia Demerouti and her colleagues have discovered that the burnout syndrome appears to have two separate but correlated components – job demands and job resources. Job demands, these authors explain, involve physical workload, time pressure, recipient contact, physical environment and shiftwork. In a non-human services workplace, emotional exhaustion is often the product of "extreme job demands" (Demerouti et al., 2001). Extreme job demands, they explain, may lead to exhaustion but not necessarily a disengagement from the workplace.

In contrast, in the matter of job resources, worker disengagement is not caused by exhaustion but by the lack of the appropriate resources (feedback, rewards, job control, participation, job security, supervisor support) required to meet the demands of the workplace. Here, an emotional withdrawal takes place resulting in disengagement and depersonalization (Demerouti et al., 2001).

The above are just some of the ongoing efforts being made to pin down what burnout is and why it occurs. Additionally, just as researchers have sought to define burnout and to determine the groups of individuals to whom it might apply, they have also tried to understand how the burnout syndrome differs from low morale, depression and stress – three other problems which plague individuals in and away from the workplace. What follows is a brief discussion of how burnout may be differentiated from low morale, depression and stress, and why.

Burnout and low morale

Morale has been defined as "the collective spirit and motivation of a group of employees" (Helms, 2006). It is

> how employees actually feel about themselves as workers, their work, their managers, their work environment, and their overall work life. It incorporates all the mental and emotional feelings, beliefs, and attitudes that individuals and groups hold regarding their job. (Helms, 2006)

Both individuals and groups of workers can suffer from low morale; however, the term is used most often to refer to the state of mind of a group of workers (Lee, 2012). Burnout, in contrast, is not a group phenomenon (although more than one worker at a time in a workplace might be burned out). The syndrome is not a shared experience but, instead, endured alone.

Not surprisingly, the symptoms exhibited by workers suffering from low morale and burnout do at times appear to overlap, particularly in the case of absenteeism and lack of enthusiasm ("10 warning signs," 2010). Low morale, however, has to do largely with job dissatisfaction and manifests itself in group attitude and behaviour, i.e., employees lack enthusiasm, discipline and commitment, appear uncooperative, complain frequently, and so on.

Individuals who are burned out, in contrast, experience a profound fatigue and lack of energy accompanied by a variety of other mental and physical problems including headaches, trouble sleeping, and feelings of helplessness and anger (Hoopes, 2006). Also important is that low morale in the workplace can be repaired with relative ease and speed given a change in supervision or management or given other necessary changes to the conditions of work. Burnout, however, takes an individual longer to recover from, even when key changes are made to the workplace.

Burnout and depression

The initial problem with the use of the term burnout, Maslach has noted, was the accusation by some in the research community that it was merely a "new label" slapped over old problems such as stress, depression, apathy, and a general unhappiness with one's employment (Maslach, 2011). But as Maslach and others explain, while certain symptoms may be shared by one or more of those mental states, both burnout and depression contain unique features which differentiate each from the other.

The two terms are sometimes used interchangeably in popular literature, but they are not synonymous. The confusion is, in part, attributable to the fact that depression, like burnout, can sometimes be defined in terms of a depletion of energy, withdrawal from other people, and feelings of reduced personal accomplishment. Additionally, just as depression may take several forms – "neurotic versus psychotic depression, agitated versus retarded depression, bipolar depression, atypical depression, dysthymia" (Farber, 2000) – several different subtypes of burnout may also be distinguished, as some researchers have suggested:

> (1) a "wornout" subtype, wherein an individual essentially gives up, or performs work in a perfunctory manner, when confronted with too much stress and too little gratification;
>
> (2) a "classic" (or "frenetic") burnout, wherein an individual works increasingly hard, to the point of exhaustion, in pursuit of sufficient gratification or accomplishment to match the extent of stress experienced; and
>
> (3) an "underchallenged" subtype of burnout, wherein an individual is faced not with an excessive degree of stress *per se* (i.e., work overload), but rather with monotonous and unstimulating work conditions that fail to provide sufficient rewards. (Farber, 2000, p. 677)

So while the lines between burnout and depression may be blurred at times, research has shown that they are distinct (Leiter and Durup, 1994;

Bakker et al., 2000) and yet consistently correlated with one another (Firth et al., 1986; Maslach and Jackson, 1986; Meier, 1984). In other words, there is a proven relationship or connection between them.

Where they differ lies in the fact that burnout is context specific; in other words, it is the product of the workplace, and the questions which are asked of members of occupational groups studied have to do with the effect of the job on the survey respondent (for example, "I feel that *this job* is hardening me emotionally," "I feel exhilarated after *working* closely with *my recipients*" (emphasis added by Leiter and Durup, 1994 citing Maslach and Jackson, 1986). Burnout originated and still is most often associated with occupational groups in which there is interaction with recipients and is the emotional toll paid by a worker as the result of that interaction.

Depression, in contrast, is about personal thoughts and feelings, and while it can be situational, it is not necessarily focused on the workplace. As a result, tools which measure depression, such as the Beck Depression Inventory (BDI (first published in 1961) and BDI-II (the 1996 revision)) do not concern themselves with the context for depression, but instead measure more broadly feelings and emotions and the physical manifestations of depression. The inventory measures, for example, the extent to which an individual is critical of herself, and the extent to which that individual feels pessimistic, guilty, worthless, or suicidal. The inventory also measures the physical manifestations of depression including crying, irritability, lack of interest in people, lack of interest in sex, loss of libido, inability to make decisions, difficulty sleeping, and poor appetite.

Individuals who are burned out, though they may lack energy, do not suffer from the same types of physical symptoms and, compared to those who are depressed, sometimes still find moments of enjoyment in life. Also, while burned out individuals are mentally and emotionally exhausted, they do not share the same inherent self-loathing, feelings of guilt, and thoughts of suicide, nor do they suffer from the same physical symptoms as those who are depressed. Typically, individuals who become burned out have been, until then, relatively mentally healthy and have not suffered from any prior psychopathology (Maslach et al., 2001). Research has shown that those who are burned out make fewer work-related errors than those who are depressed (Fahrenkopf et al., 2008) and any reduction in their work performance is generally the result of negative attitudes and behaviours (Maslach et al., 2001).

Finally, while those who are burned out and those who are depressed might have problems sleeping, the types of problems experienced by each group in this regard also differ measurably (Brenninkmeyer et al., 2001; Hoogduin et al., 1996). The same is true regarding feelings of self-worth; depressed individuals are more likely to assess themselves lower than those who are burned out in terms of "superiority" (i.e., defined as inferior versus superior rank, undesirable versus desirable appearance, and unaccepted versus accepted by others). Depressed individuals, when questioned on these items, are more likely to respond as though they have given up; only the most severely burned out individuals who feel utterly defeated have been shown to respond in a similar manner (Brenninkmeyer et al., 2001).

Burnout and stress

Researchers often speak of burnout as "an extension of the construct of stress" or a "distinct consequence of work-related stress" (Halbesleben and Buckley, 2004, p. 871). Unlike stress, however, which can occur in any part of an individual's life, burnout is most often related specifically to the workplace. Brill (1984), for example, defines burnout as an

> expectationally mediated, job-related, dysphoric and dysfunctional state in an individual without major psychopathology who has (1) functioned for a time at adequate performance and affectual levels in the same job situation and who (2) will not recover to previous levels without outside help or environmental rearrangement. (p. 15)

Burnout is the point at which employees can no longer deal with the demands made of them, accompanied by a loss of meaning in that they no longer believe in the work they are doing (Casserley and Megginson, 2009). As Pines and Aronson (1988) explain,

> [p]eople experience burnout as a gradual erosion of their spirit and zest as a result of the daily struggles and chronic stresses that are typical of everyday life and work – too many pressures, conflicts, demands, and too few emotional rewards, acknowledgements, and successes. Stress, in and of itself, does not cause burnout. People are often able to flourish in stressful, demanding careers if they feel valuable and appreciated and that their work has significance. They burn out when their work has no meaning and stress continuously outweighs support and rewards. (p. 11)

Individuals who are stressed are often over-engaged, whereas those who are burned out are disengaged. Stress is characterized by an emotional state which is over-reactive, producing a sense of urgency and hyperactivity. In contrast, the burned out individual has blunted emotions and feels helpless, detached and defeated (Smith et al., 2012). Individuals who are stressed and/or burning out can feel that "control of the boundaries between work and non-work [is] temporarily threatened." Even so, when periods of stress are not too prolonged most workers believe they have the resources and the ability "sufficient to wrest control back again" (Casserley and Megginson, 2009, p. 16). Those who suffer from burnout no longer have that same confidence.

Unlike burnout, stress can be short term and buffered by adequate organizational infrastructure (Brill, 1984). Also, unlike burnout, not all stress is bad. Job stress which is short term is not unusual for most occupations, and some employees have been shown to thrive in a stressful environment (Pithers, 1995; Whitehead and Ryba, 1995). Chronic and long term stress, however, can lead to burnout. Also, while stress may be here today and gone tomorrow, burnout is something which develops over time (Leiter, 1993; Maslach and Leiter, 1997; Schaufeli and Enzmann, 1998) and from which it takes time to recover.

Expectations, however, play a greater role in burnout, which is, in part, a response to the difference in expectations between what the job will be versus what it really is. Burnout can also be fed by unrealistic expectations of oneself. As one researcher has noted, "[b]urnout is considered pre-eminently a disorder of the over-committed and thwarted" (Roberts, 1997).

Using the terms stress and burnout interchangeably is misleading in that it fails to account for the

> strong phenomenological perception of "inconsequentiality" among those who feel burned – the sense that their efforts on their job are not met with commensurate rewards, satisfactions, or fulfillment. (Farber, 2000)

Burnout has much to do with stress, but ultimately has hallmarks that are not necessarily present in other stress-related disorders.

Classification

The term burnout does not appear in the fourth and latest edition of *The Diagnostic and Statistical Manual of Mental Disorders* (DSM-IV) published in 1994. This manual, which classifies mental disorders, is published by the American Psychiatric Association and used primarily in North America. At the time of writing, the next major revision ("DSM-V") was due to be published in 2013, although whether the term will appear in that edition is yet unknown.

One of the very practical ramifications of the term not appearing in the DSM-IV is that treatment is often not supported by insurance companies, and there has not been the same kind of research involving diagnosis, treatment and effective therapies as in those parts of the world which regard it as a recognized disorder. Instead, in the regions which use the DSM-IV, burnout is generally treated as "a nondistinct stress-related phenomenon. Thus, suggestions for its remediation tend to parallel those for any stress-related disorder" (Farber, 2000, p. 676).

In contrast to the North American situation, the *International Statistical Classification of Diseases and Related Health Problems* (ICD-10) is the classification work used primarily in Europe and in certain other parts of the world. The ICD-10, published by the World Health Organization, is a manual which classifies diseases and other health-related problems that appear on death certificates and in medical records. World Health Organization member states rely on it to help generate meaningful statistics and related information on health and mortality in various segments of the population. The term burnout has appeared in this manual since 1992. It is classified under the heading "Problems related to life management difficulty" (Z73) and is defined as a "State of vital exhaustion" (ICD-10, 2012).

Accordingly, in certain European countries (e.g., Sweden and the Netherlands), burnout is a legitimate medical diagnosis, and individuals who are diagnosed as burned out by a psychiatrist are eligible for sick benefits or disability. The criteria for the diagnosis are:

> physiological or mental symptoms of exhaustion for at least two weeks, an essential lack of psychological energy, and symptoms such as difficulties to concentrate, decreased ability to cope with stress, irritability or emotional instability, sleep disturbances, muscle pain, dizziness or palpitations. These symptoms have to occur every day during a two-week period and must cause significant suffering with impaired work capacity. Finally the symptoms must not be related to other psychiatric diagnosis, substance abuse, or medical diagnosis. (Schaufeli et al., 2009, pp. 213–14)

Health care professionals in these European countries are not only able to diagnose burnout, but are also able to provide forms of treatment and other interventions. In various places of employment, workshops are held on preventing and dealing with burnout. Health care professionals also act as consultants to a variety of organizations on the problem of burnout (Schaufeli et al., 2009).

Outside of these specific countries, however, burnout remains a vaguer malady. While the term appears with relative frequency in the news as a catch-all word to define overwork and exhaustion, the specifics of the syndrome and how to deal with it are still mysteries to most employees and their organizations. What we turn to next is a discussion of some of these specifics.

Symptoms and diagnosis

Individuals who burn out are likely to blame themselves, thinking that something must be wrong with them if they are unable to handle the stress (Maslach, 2003). This is often reinforced by the situation in which they find themselves. Chronically stressed individuals remember a time when the stress wasn't a problem for them. As one employee commented,

> *I pretty much felt I was untouchable and I could handle anything. And in point of fact the jobs that I was given to do were the most difficult, and I thrived on it. I was given impossible people to work with and I could get results that no one else could get and that fed the sense of superiority that I was invincible.* (Glouberman, 2002, p. 68; italic in original)

Burned out workers often don't understand that burnout is the end result of prolonged stress experienced on a daily basis, and that it has taken some time for the cumulative effects to manifest themselves. They are unaware that they have been burning out over a period of time; they only know that they once were able to function more effectively than they do now; hence the tendency to blame themselves or others (co-workers, supervisors, clients) with whom they work.

Given that burnout takes place over time, one of the great difficulties lies in identifying it as it is taking place. Burnout has been described as a "fuzzy set" (Burisch, 1993), which means that it is not easy to distinguish its symptoms from those of similar phenomena such as low morale, depression, exhaustion and stress. Nonetheless, burnout is a syndrome which is empirically distinguishable from these other similar phenomena and "not just a 'flabby' synonym for more established symptomatic states" (Roberts, 1997, p. 282).

As we have seen, burnout shares certain traits with other problems such as stress and depression, yet it is still possible to identify burnout as a distinct problem. What follows is a discussion of some of the symptoms that have been definitively identified in the literature, including changes in behavior, changes in feelings, changes in thinking and changes in health.

a) Changes in behavior

One of the hallmarks of burnout is withdrawal from others. As Christina Maslach notes, "[o]ne way people try to get out from under their emotional burden is by cutting back on their involvement with others. They want to reduce their contact with people to the bare minimum required to get the job done" (Maslach, 1982, p. 3). Co-workers and supervisors tend to notice this. Indeed, studies in which co-workers describe the behavior of their colleagues in the months leading up to complete burnout note this withdrawal from co-workers and a tendency to want to complete tasks alone (Ericson-Lidman and Strandberg, 2007).

Employees who are burning out distance themselves from an assortment of people: supervisors, co-workers and/or the group of people whom they are trained to serve or help. Where these employees once had a great interest in the welfare of the recipients of their services, for example, they begin treating these same recipients with callousness and negativity (Maslach et al., 2001).

They spend less time and energy on their work and are absent more often. Once an individual has burned out, both the quality and the quantity of the work may suffer significantly.

Withdrawal can also include a withdrawal from social activities. Individuals who are burning out begin to ignore or cancel invitations and do not pay attention to relationships with those who are close to them. Even so, they often still believe the situation is temporary (Ekstedt and Fagerberg, 2005).

Other unexpected changes in behavior appear. Some individuals report that, in the course of burning out, they have found themselves losing their temper in a way that was unusual and quite unlike their "normal" selves (Casserley and Megginson, 2009). Emotionally exhausted workers suddenly become more irritable or hostile, and exchanges with co-workers become emotionally charged. As one burned out worker described it,

> *The person sitting there wasn't me. I wasn't really there. I wasn't enjoying anything. I felt like I didn't fit my life. I was in the wrong position, doing the wrong thing, and I shouldn't be there.*
>
> *I was definitely losing it at work. I started to feel more indifferent to my clients. I felt angry at colleagues. I had a combination of anger, cynicism, hate and disillusionment. I was furious – hating myself and hating them at the same time. An impotent rage. I was also terribly exhausted, and started getting physical symptoms. I also began to get more addicted to things, one minute chocolate, then drinking, then a phase of casual sex.* (Glouberman, 2002, pp. 56–7; italic in original)

The particular changes in behavior like those described above are among the hallmarks of burnout. Not only are co-workers and recipients surprised by these changes, but the burned out employees in question no longer recognize their old selves.

b) Changes in feelings

Individuals who are burned out no longer feel the same way they did about the people they were hired to help, or about the organization for which they work. They are often, at first, frustrated and angry; later they become disengaged from the workplace and are no longer enthusiastic about the work they once enjoyed. According to one sufferer,

> *I was just fed up with living. That was the main symptom. Everything in life was an effort. There was no joy in anything I was doing. I couldn't feel anything and wondered if I had any feelings at all. I think I'd been pretty shut down for quite a lot of my life, but I was just absolutely emotionally dead. I felt like I'd lost all sense of who I was and what I was.* (Gloubermann, 2002, p. 149; italic in original)

Schaufeli and Enzmann (1998) discuss the affective, cognitive, physical, behavioral and motivational signals which indicate burnout. Significantly, the affective, cognitive and motivational signals all have to do with changes in feelings. For example, affective signals include depressive/changing moods, tearfulness and exhaustion. Cognitive signals include feelings of helplessness, hopelessness, powerlessness, guilt and feeling trapped: work has lost its meaning. Finally, motivational signals include loss of enthusiasm and zeal, along with disappointment and resignation (Schaufeli and Enzmann, 1998).

Along similar lines, there are noticeable changes in attitude which are largely negative. Burned out employees no longer appear to have the work goals they once had; not only have they lost their initial idealism, but they may also display a new heightened self-interest which is yet another means by which they are attempting to cope (Benbow, 1998).

c) Changes in thinking

Employees who are burning out worry about being able to fulfill their work responsibilities, in part, because they are having difficulty concentrating and making decisions (Ekstedt and Fagerberg, 2005). These individuals describe problems with memory and "blackouts" during which "they suddenly could not remember how to perform familiar tasks, or [they forget] to carry out routine work, which then took a great deal of time to catch up on" (Ekstedt and Fagerberg, 2005, p. 64). One employee explained,

> *I went into a kind of blank space, where I was almost frozen into inaction and beyond caring what happened. Some days it was a heavy feeling, like wading through treacle and everything being an enormous effort, or like having your foot on the accelerator and the brake at the same time so that the net response was zero. Other days it just felt remote and disconnected and empty/indifferent.*
>
> *I still more or less functioned day to day, in a going-through the motions kind of way, but the cracks began to show in other silly but completely uncharacteristic ways: things like going out and forgetting to turn the iron off, or realizing only at the very end of a day that I'd put on the jacket of one suit with the skirt of another that didn't go at all, and on a couple of occasions, I lost my temper with different bosses. I knew I couldn't go on like this, but couldn't see an alternative.* (Glouberman, 2002, p. 149; italic in original)

Such employees no longer have a sense of accomplishment in their job. While there is some debate as to whether this is a result of the other two dimensions of burnout (exhaustion and depersonalization/cynicism) (Byrne, 1994; Lee and Ashforth 1996) or whether reduced personal accomplishment develops in parallel with the other two aspects (Leiter, 1993), this conviction that one is not working as effectively, competently or productively is a hallmark of burnout.

Paradoxically, despite their worries about job performance, studies have shown that, on the road to burnout, employees' perception of their own work performance differs from how it is *objectively* viewed. For example, when Dutch nurses were asked to rate their job performance, researchers discovered that burnout was positively related to how well one is doing on the job. In other words, those individuals who are burning out feel they are performing poorly when in fact, their supervisors noticed higher performance levels (Keijsers et al., 1995; Parker and Kulik, 1995).

Nonetheless, colleagues do notice the discrepancy between how their at-risk colleagues feel they are doing at work and how they are actually doing. For example, one study in which co-workers were asked to describe the behavior of colleagues in the months before those colleagues burned out reported that their at-risk colleagues were working harder to exacting standards but often felt that they were not doing a good job. The colleagues on the verge of burnout believed they could not perform their work adequately because they felt emotionally drained and stretched for time (Ericson-Lidman and Strandberg, 2007).

Eventually, however, the fear that one is not doing a good job becomes a self-fulfilling prophecy. Once burned out, workers really do have difficulty performing their work, often because of increased absenteeism which may, in its most severe form, eventually lead to a change of jobs (Toppinen-Tanner et al., 2005).

d) Changes in health

Physically, burnout manifests itself most profoundly as constant fatigue. Individuals feel drained and exhausted; they suffer from sleep disturbances and insomnia. Here is one self-analysis:

> *Exhaustion. I was driving myself to carry on and to work harder and harder and I was having to work harder to achieve the same or less. There were huge amounts of effort put into not achieving much, but I was pushing myself very hard. I felt exhausted and just couldn't carry on. I was disconnected from people. I started suffering from quite bad insomnia. I wasn't sleeping more than two or three hours a night.* (Glouberman, 2002, p. 148; italic in original)

Interestingly, disturbances in sleep are so much a part of the burnout syndrome that one can predict the other or vice versa. Researchers, for example, have shown that not only does burnout predict subsequent cases of insomnia or changes in intensity of insomnia but that insomnia or a change in intensity of insomnia predicts subsequent cases of burnout and changes in levels of burnout (Armon, 2008). What all this means is that employees at risk have trouble sleeping. They also find themselves unable to unwind in the evenings (Ekstedt and Fagerberg, 2005).

A century ago, the renowned physician Sir William Osler believed "stress and strain" along with hard work and worry were the preconditions of heart disease (Spielberger, 1979). Since that time, burnout has been linked to a number of health-related problems, including an increased risk of cardiovascular disease (Melamed et al., 2006). It has also been associated with diabetes (Melamed et al., 2006) and musculoskeletal disorders (Honkonen et al., 2006).

Those burning out often find themselves suffering from headaches and muscle tension or pain in the neck, arms and shoulders (Ekstedt and Fagerberg, 2005), high blood pressure, lingering colds and gastrointestinal problems (Freudenberger, 1974; Maslach, 1976).To this list, researchers have also added "dizziness, nausea, allergies, breathing difficulties, skin problems, muscle aches and stiffness, menstrual difficulties, swollen glands, sore throat, recurrent flu, infections, colds, headaches, digestive problems, and back pain" (Brock and Grady, 2000, p. 30).

Individuals who are burning out may also attempt to cope in ways that are self-destructive. It is not unusual to find that they are self-medicating with alcohol and drugs which tend to exacerbate the problems. Ultimately, these symptoms and destructive coping mechanisms spill over into their private lives affecting relationships with family and friends – the very individuals needed more than ever during these difficult times.

Conclusion

Burnout is a syndrome which was originally thought to affect only human services professionals but is now understood to affect a much broader range of workers. Its primary characteristic is emotional exhaustion, although there is a wide consensus among researchers that it is accompanied by feelings of depersonalization/cynicism and feelings of reduced personal accomplishment/ineffectiveness.

Burnout shares some characteristics with low morale, stress and depression, but is unique and stands apart from these other phenomena. Specific changes, for example, in the way an employee thinks, behaves, and feels can signal that the individual in question is at risk of burnout. Specific health problems can also accompany the syndrome.

The co-workers of employees who have burned out have reported behaviors which, in retrospect, were clear signs that their co-worker was in distress. These behaviors include the desire to work alone; the appearance of being weighed down by the demands of the job; withdrawal from co-workers, service recipients, and work; isolation; distraction; sleep disturbances; and the appearance of being at the breaking point (Ericson-Lidman and Strandberg, 2007). These are acknowledged to be the classic signs of burning out but they provide no information, in themselves, as to why they are occurring in the first place. In our next chapter, we explore some of the key socio-demographic and workplace factors which researchers have determined put workers at greater risk for burnout.

References

10 warning signs of low morale . . . and 7 ways to boost it. (2010) *HR Specialist*, 8(6), 6.

American Psychiatric Association. (1994) *Diagnostic and Statistical Manual of Mental Disorders: DSM-IV*. Washington, DC: American Psychiatric Association.

Armon, G., Shirom, A., Shapira, I., and Melamed, S. (2008) On the nature of burnout–insomnia relationships: A prospective study of employed adults. *Journal of Psychosomatic Research*, 65(1), 5–12.

Awa, W.L., Plaumann, M., and Walter, U. (2010) Burnout prevention: A review of intervention programs. *Patient Education and Counselling*, 78(2), 184–90.

Bakker, A.B., Schaufeli, W.B., Demerouti, E., Janssen, P., Van Der Hulst, R., and Brouwer, J. (2000) Using equity theory to examine the difference between burnout and depression. *Anxiety, Stress & Coping*, 13(3), 247–68.

Beck, A.T., Ward, C.H., Mendelson, M., Mock, J., and Erbaugh, J. (1961) An inventory for measuring depression. *Archives of General Psychiatry*, 4, 561–71.

Beck, A.T., Steer, R.A., and Brown, G.K. (1996) *Manual for the Beck Depression Inventory-II*. San Antonio, Texas: Psychological Corporation.

Beemsterboer, J., and Baum, B. (1984) "Burnout": Definitions and health care management. *Social Work in Health Care*, 70(1), 97–109.

Benbow, S. (1998) Burnout: Current knowledge and relevance to old age psychiatry. *International Journal of Geriatric Psychiatry*, 13(8), 520–6.

Bradley, H.B. (1969) Designing for change: Problems of planned innovation in corrections. *Annals of the American Academy of Political and Social Science*, 381(1), 89–98.

Brenninkmeyer, V., Van Yperen, N.W., and Buunk, B.P. (2001) Burnout and depression are not identical twins: Is decline of superiority a distinguishing feature? *Personality and Individual Differences*, 30(5), 873–80.

Brill, P. (1984) The need for an operational definition of burnout. *Family & Community Health*, 6(4), 12–24.

Brock, B.L., and Grady, M.L. (2000) *Rekindling the Flame: Principals Combating Teacher Burnout.* Thousand Oaks, California: Corwin Press.

Burisch, M. (1993) In search of theory: Some ruminations on the nature and etiology of burnout. In W. Schaufeli, C. Maslach, and T. Marek (Eds.), *Professional Burnout: Recent Developments in Theory and Research.* Washington, DC: Taylor and Francis.

Byrne, B.M. (1994) Burnout: Testing for the validity, replication, and invariance of causal structure across elementary, intermediate and secondary teachers. *American Education Research Journal*, 31(3), 645–73.

Casserley, T.B., and Megginson, D. (2009) *Learning from Burnout: Developing Sustainable Leaders and Avoiding Career Derailment.* Oxford, UK: Butterworth-Heinemann.

Contenta, S. (23 January 2010) "21st century syndrome" or just a case of the blahs? An intriguing diagnosis posits that we live in an age so stressful we can no longer cope. *Toronto Star*, IN01.

Demerouti, E., Bakker, A.B., Nachreiner, F., and Schaufeli, W.B. (2001) The job demands–resources model of burnout. *Journal of Applied Psychology*, 86(3), 499–512.

Ekstedt, M., and Fagerberg, I. (2005) Lived experiences of the time preceding burnout. *Journal of Advanced Nursing*, 49, 59–67.

Ericson-Lidman, E., and Strandberg, G. (2007) Burnout: Co-workers' perceptions of signs preceding workmates' burnout. *Journal of Advanced Nursing*, 60(2), 199–208.

Fahrenkopf, A.M., Sectish, T.C., Barger, L.K., Sharek, P.J., Lewin, D., Chiang, V.W., Edwards, S., Wiedermann, B.L., and Landrigan, C.P. (2008) Rates of medication errors among depressed and burnt out residents: Prospective cohort study. *British Medical Journal*, 336(7642), 488–91.

Farber, B.A. (2000) Treatment strategies for different types of teacher burnout. *Journal of Clinical Psychology*, 56(5), 675–89.

Firth, H., McIntee, J., McKeown, P., and Britton, P. (1986) Burnout and professional depression: Related concepts? *Journal of Advanced Nursing*, 11, 633–41.

Freudenberger, H. (1974) Staff burn-out. *Journal of Social Issues*, 30, 159–65.

Freudenberger, H. (1980) *Burnout: The High Cost of High Achievement.* Garden City, New York: Anchor Press.

Glouberman, D. (2002) *The Joy of Burnout.* Isle of Wight: Skyros Books.

Gray, R. (3 April 2005) Is stress all in our heads? *Scotland on Sunday*, 13.

Greene, G. (1977) *A Burnt-out Case.* Harmondsworth, UK: Penguin. (Original work published 1960.)

Halbesleben, J.R.B., and Buckley, M.R. (2004) Burnout in organizational life. *Journal of Management*, 30(6), 859–79.

Health and Safety Executive (2011) *Stress and psychological disorders.* Retrieved from *http://www.hse.gov.uk/statistics/causdis/stress/stress.pdf.*

Helms, M.M. (Ed.) (2006) Morale. *Encyclopedia of Management.* Detroit, Michigan: Gale.

Honkonen, T., Ahola, K., Pertovaara, M., Isometä, E., Kalimo, R., Nykyri, E., Aromaa, A., and Lönnqvisit, J. (2006) The association between burnout and physical illness in the general population – results from the Finnish Health 2000 Study. *Journal of Psychosomatic Research*, 61(1), 59–66.

Hoogduin, C.A.L., Schaap, C.P.D.R., and Methorst, G.J. (1996) Burnout: Klinisch beeld en diagnostiek. (Burnout: Clinical picture and diagnostics). In C.A.L. Hoogduin, C.P.D.R. Schaap, A.J. Kladler, and W.A. Hoogduin. (Eds.), *Behandelingsstrategieen bij Burnout (Treatment Strategies of Burnout)*. Houten/Diegem, Netherlands: Bohn Stafleu Van Loghum.

Hoopes, B. (2006) Key burnout warning signs. *Landscape Management*, 45(8), 16.

Huge cost of stress (10 October 2010) *Sunday Telegraph*, 40.

International Statistical Classification of Diseases and Related Health Problems, 10th Revision (ICD-10), Version 2010, "Burn-out." Retrieved from *http://apps.who.int/classifications/icd10/browse/2010/en#/Z73.0.*

ICD-10 (2012) 2012 ICD-10-CM Diagnosis Codes. Retrieved from *http://www.icd10data.com/ICD10CM/Codes/Z00-Z99/Z69-Z76/Z73-/Z73.0.*

Jobs Mail: Staff burnout increase bad for business (2 June 2005) *Evening Mail*, 30.

Kalia, M. (2002) Assessing the economic impact of stress – the modern day hidden epidemic. *Metabolism, Clinical & Experimental*, 51(6) Supp. 1, 49–53.

Keijsers, G.J., Schaufeli, W.B., Le Blanc, P.M., Zwerts, C., and Miranda, D.R. (1995) Performance and burnout in intensive care units. *Work & Stress*, 9, 513–27.

Kristensen, T.S., Borritz, M., Villadsen, E., and Christensen, K.B. (2005) The Copenhagen Burnout Inventory: A new tool for the assessment of burnout. *Work & Stress*, 19(3), 192–207.

Lee, R.T. and Ashforth, B.E. (1996) A meta-analytic examination of the correlates of the dimensions of burnout. *Journal of Applied Psychology*, 81, 123–33.

Lee, T.W. (2 November 2012) Employee morale. *The Blackwell Encyclopedia of Management*. Oxford, UK: Blackwell Publishing.

Leiter, M.P. (1993) Burnout as a developmental process: Consideration of models. In W.B. Schaufeli, C. Maslach, and T. Marek (Eds.), *Professional Burnout: Recent Developments in Theory and Research* (pp. 237–50). Washington, DC: Taylor and Francis.

Leiter, M.P., and Durup, J. (1994) The discriminant validity of burnout and depression: A confirmatory factor analytic study. *Anxiety, Stress & Coping: An International Journal*, 7(4), 357–73.

Levinson, M. (2012) When more gadgets means less stress. *CIO*, 25: 8, 10.

Maslach, C. (1976) Burned out. *Human Behavior*, 5, 16–22.

Maslach, C. (1982) *Burnout: The Cost of Caring*. Englewood Cliffs, New Jersey: Prentice-Hall.

Maslach, C. (2003) Job burnout: New directions in research and intervention. *Current Directions in Psychological Science*, 12(5), 189–92.

Maslach, C. (2011) Engagement research: Some thoughts from a burnout perspective. *European Journal of Work and Organizational Psychology*, 20(1), 47–52.

Maslach, C. and Jackson, S.E. (1981) The measurement of experienced burnout. *Journal of Occupational Behaviour*, 2, 99–113.

Maslach, C. and Jackson, S.E. (1986) *Maslach Burnout Inventory* (2nd ed.). Palo Alto, California: Consulting Psychologists Press.

Maslach, C. Jackson, S.E., and Leiter, M.P. (1996) *MBI: The Maslach Burnout Inventory Manual* (3rd ed.). Palo Alto, California: Consulting Psychologists Press.

Maslach, C., and Leiter, M.P. (1997) *The Truth about Burnout: How Organizations Cause Personal Stress and What to Do about It.* San Francisco, California: Jossey-Bass.

Maslach, C., Schaufeli, W.B., and Leiter, M.P. (2001) Job burnout. In S.T. Fiske, D.L. Schachter, and C. Zahn-Waxler (Eds.), *Annual Review of Psychology*, 52, 397–422.

Meier, S.T. (1984) The construct validity of burnout. *Journal of Occupational Psychology*, 57, 211–19.

Melamed, S., Shirom, A., Toker, S., and Shapira, I. (2006) Burnout and risk of type 2 diabetes: A prospective study of apparently healthy employed persons. *Psychosomatic Medicine*, 68, 863–9.

National Institute for Occupational Safety and Health (NIOSH) (1999) *Stress at work*, Publication Number 99–101. Retrieved from *http://www.cdc.gov/niosh/docs/99-101/pdfs/99-101.pdf.*

Nauratil, M.J. (1989) *The Alienated Librarian.* New York, New York: Greenwood Press.

Pannozzo, L., and Landon, L. (2005) Overwork and underwork take heavy toll on our economy. *CCPA Monitor*, 11(10), 24.

Parker, P.A., and Kulik, J.A. (1995) Burnout, self and supervisor-related job performance, and absenteeism among nurses. *Journal of Behavioral Medicine*, 18, 581–99.

Pines, A., and Aronson, E. (1988) *Career Burnout: Causes and Cures.* New York, New York: The Free Press.

Pithers, R.T. (1995) Teacher stress research: Problems and progress. *British Journal of Educational Psychology*, 65, 387–92.

Roberts, G.A. (1997) Prevention of burn-out. *Advances in Psychiatric Treatment*, 3, 282–9.

Schaufeli, W., and Enzmann, D. (1998) *The Burnout Companion to Study and Practice: A Critical Analysis.* London, UK: Taylor and Francis.

Schaufeli, W.B., and Greenglass, E.R. (2001) Introduction to special issue on burnout and health. *Psychology and Health*, 16, 501–10.

Schaufeli, W.B., Leiter, M.P., and Maslach, C. (2009) Burnout: 35 years of research and practice. *Career Development International*, 14(3), 204–20.

Senior, J. (26 November 2006) Can't get no satisfaction: In a culture where work can be a religion, burnout is its crisis of faith. Retrieved from *http://nymag.com/news/features/24757/.*

Shirom, A. (1989) Burnout in work organizations. In C.L. Cooper and I. Robertson (Eds.), *International Review of Industrial and Organizational Psychology* (pp. 26–48). Chichester, UK: Wiley.

Smith, M., Segal, J., and Segal, R. (2012) *Preventing Burnout: Signs, Symptoms, Causes, and Coping Strategies.* Retrieved from *http://www.helpguide.org/mental/burnout_signs_symptoms.htm.*

Spielberger, C. (1979) *Understanding Stress and Anxiety.* New York, New York: Harper and Row.

Study, E. (9 August 2012) Gadgets affecting work/life balance. *Dayton Daily News*, OZWR.5.

Toppinen-Tanner, S., Ojajärvi, A., Väänaänen, A., Kalimo, R., and Jäppinen, P. (2005) Burnout as a predictor of medically certified sick-leave absences and their diagnosed causes. *Behavioral Medicine*, 31(1), 18–32.

Walter, L. (2012) The break-room: Stressing out. Retrieved from *http://ehstoday.com/health/break-room-stressing-out-0302/*.

Whitehead, A.J., and Ryba, K. (1995) New Zealand teachers' perceptions of occupational stress and coping strategies. *New Zealand Journal of Educational Studies*, 30, 177–88.

2

Factors contributing to burnout

Abstract: This chapter deals with the individual and socio-demographic factors which play a role in burnout. These factors include age or years of experience, personality, locus of control, marital status, gender, work–home interference and expectations. Environmental and organizational factors which could lead to burnout are also discussed including workload, underwork, type of work, physical environment, conflict, role conflict and role ambiguity, control, social support, reciprocity, social comparison and leadership style.

Key words: burnout, age and experience, personality, gender, work–home interference, expectations, workload and type of work, physical environment, conflict and support.

Introduction

People do not generally burn out in response to one stressful factor in the workplace but rather to a number of factors working in concert (Gabris and Ihrke, 1996). But which factors? Since Freudenberger and Maslach first began writing about the syndrome, researchers have sought to isolate the workplace conditions responsible for burnout, and many of these conditions have been identified and confirmed by empirical studies over the years. For example, it is now widely recognized that workload (quantitative and qualitative) correlates positively with burnout (Maslach and Leiter, 1997), as do excessive job demands (Burisch, 2006).

Yet, despite such matters upon which there is agreement among researchers, there is much we still do not know. A number of studies, for instance, have provided us with mixed results in terms of whether certain socio-demographic factors might be responsible for burnout. Some researchers (Dillon and Tanner, 1995; Yildirim, 2008) have found no relation between the dimensions of gender, age or marital status and burnout, while others have determined that the relationship between these very dimensions is significant (Jackson, 1993; Cordes and Dougherty, 1993). Results such as these are confusing, to say the least, and remind us that more research is necessary to determine if, indeed, certain socio-demographic factors are significant, but only perhaps amongst certain groups or types of workers.

What follows is a brief overview of the various potential causes of burnout which have been studied and the extent to which these have been linked to the syndrome. Individual and socio-demographic factors are discussed first; this will be followed by a consideration of those workplace conditions which researchers believe put individuals at greater risk for burnout.

Individual and socio-demographic factors

Age or years of experience

Researchers have long attempted to determine whether age is a significant factor in employee burnout. Their findings are, however, mixed. Some studies have shown that age is significantly correlated with the three dimensions of burnout (Duffy et al., 2009), and that younger workers are more likely than their older counterparts to feel emotionally exhausted and to experience depersonalization (Williams, 1989). Why is this?

One theory is that it is common for young people to enter certain professions or lines of work for which they are not suited. This lack of "fit" is likely to lead to higher rates of stress and burnout resolved only when these individuals leave the field to look for another line of work. This "natural selection" implies that older workers who remain in the job or profession are more temperamentally suited to the job or are likely to have developed successful coping skills and strategies along the way (Brewer and Shapard, 2004).

A second theory to explain the difference in burnout scores between younger and older employees amongst certain groups of workers has to do with the amount of training or education required to do the job. Some workers are older than others when they enter the workplace. As a result, they have had more time to understand the pros and cons of their chosen profession, and have the maturity to handle better certain stressors.

Individuals who enter the teaching profession, for instance, are generally younger when they start working than individuals who begin to practice medicine. The latter are more mature simply because it takes many more years of schooling and training to qualify as a physician. Hence the theory that, in those areas of employment in which professionals start working at a younger age, they are more likely to show higher rates of burnout than in the areas of employment in which professionals starting their careers are older (Brewer and Shapard, 2004).

Nonetheless, neither of the above theories accounts for the results of some studies which indicate no difference in burnout scores between younger and older employees (Konert, 1997; Ahola et al., 2008), or which indicate that burnout does not decrease significantly with age (Schwartz et al., 2007). Clearly, more research is needed to determine when and why age comes up as a significant factor.

If experience and age are two variables which are not always easily parcelled out, there also exists some uncertainty as to whether the number of years of experience at the job has any bearing on being at greater or lesser risk for burnout. Various studies have indeed suggested that workers who have not been on the job for a long time are more likely to experience burnout (Bennett and Michie, 1991; Kantas and Vassilaki, 1997; Brewer and Shapard, 2004).

These results, however, are not universally shared. Other studies (Gaines and Jermier, 1983; Ray and Miller, 1994; Haddad, 1998) have found no difference in levels of burnout between workers who were new to the job and those who had many more years of experience. Once again, the research raises as many questions as it answers and invites more investigation to determine how reliable age and years of experience are as predictors of burnout.

Personality

Workplace characteristics, far more than personality traits, are generally found to be the best predictors of burnout (Barber and Iwai, 1996; Thomsen et al., 1999). As Maslach and Leiter (1997) assert,

> research and consulting work provide clear and consistent evidence that the roots of burnout stretch far beyond the individual and into the work environment. Burnout does not result from a genetic predisposition to grumpiness, a depressive personality, or general weakness. It is not caused by a failure of character or a lack of

> ambition. It is not a personality defect or a clinical syndrome. It is an occupational problem. (p. 34)

Yet despite such claims, theories do abound as to the types of individuals who are likely to burn out. These are said to include workaholics, enthusiastic high-achievers, and "type-A, driven, 'stress junkie[s]' " (Roberts, 1997). As Glouberman (2002) writes,

> [t]hose of us prone to burnout tend to start out as high-energy, ambitious and capable achievers. Whatever our calling in life, we manage to be the people who are enthusiastic, work hard and do whatever needs doing no matter what the cost. We often perceive ourselves as holding together situations which would fall apart without us. Sometimes this is true. Many of us think of ourselves as unlimited in our energy, even as "Supermen" or "Superwomen". This is less a matter of pride than a statement of how our life requires us to be. (p. 55)

Not everyone agrees, however, that enthusiastic over-achievers are most at risk. Some research indicates that employees who are sensitive and empathetic, or anxious and obsessive (Burke and Richardson, 1996) are more likely to burn out. The syndrome is also thought to occur more frequently among those employees who have unrealistic expectations about their career and accomplishments (Casserley and Megginson, 2009).

Empirical work has been able to show that, in the human services professions, perfectionists are particularly at risk. Individuals who score higher on perfectionism are those who generally experience greater stress levels, and those stress levels, in turn, lead to a greater likelihood of burnout (D'Souza et al., 2011).

As well, a study conducted by Casserley and Megginson (2009) found that individuals who derive a large part of their identity through their work and career and look for accolades or recognition in return are more liable to burn out. Casserley and Megginson interviewed 105 "high flyers" from the "high potential talent pool" at four different organizations. They discovered that employees who burn out set no boundaries between the job and home; personal sacrifices are viewed as part of the job as is the sacrifice of various physical and emotional needs.

Casserley and Megginson's "high flyer" study revealed that these same employees have a poorly developed reflexive ability, i.e., they just respond to external stimuli, but lack the ability to discuss, evaluate, and interpret and thereby take control and make choices. Individuals who eventually suffered from burnout had the benefit of hindsight to help them understand what they experienced. When asked about what they were thinking as they were burning out, many said that they had focused only on the immediate problems of the job, finding themselves too exhausted or numbed to do much more.

In addition to diagnosing a poorly developed reflexive ability in burnout sufferers, researchers have also established a connection between burnout and neuroticism, as well as burnout and introversion (Fontana and Abouserie, 1993; Mills and Huebner, 1998; Zellars et al., 2000). Neuroticism is a personality trait; the term is used to describe individuals who "express more negative emotions, emotional instability and stress reaction and, therefore, they become more vulnerable to both burnout and the majority of psychopathological disturbances" (Cano-García, Padilla-Muñoz, and Carrasco-Ortiz, 2005, p. 937; Watson et al., 1994). Neurotic individuals are prone to worry, nervousness, and sadness. Because they are temperamental, they are less able to cope effectively with stress and are more prone to burnout. In contrast, those who are more emotionally stable are able to adapt more effectively to stressful situations.

In contrast to neurotic individuals, those who are introverted are described as passive, having a "lack of interest in social exchanges and less disposition towards positive emotionality; all of these being characteristics that foster emotional exhaustion and depersonalization while they diminish personal accomplishment" (Cano-García, Padilla-Muñoz, and Carrasco-Ortiz, 2005, p. 937). On the other hand, as one might expect, studies have shown that those who are extraverted, emotionally stable, open to experience, and agreeable are less likely to experience depersonalization, or to feel diminished personal accomplishment (Ghorpade et al., 2011).

In addition to their findings on the above personality traits, several studies have looked at "hardiness" as a psychological trait, and particularly the extent to which hardy individuals are less likely to experience burnout. Hardiness has three dimensions which have been observed: hardy individuals feel they are in control of situations; they are committed (i.e., involved in activities and curious about the world around them); and they are open to the challenges and opportunities presented by constant change in their lives. This attribute of hardiness is said to allow individuals to cope with and successfully resolve stressful situations; they are also less likely to view problems as personally threatening.

But on this matter of the extent to which hardiness buffers the effects of stress, research results have been inconsistent. Certain studies have found that hardiness does act as a buffer for men but plays no role whatever in helping women cope with workplace stress (Klag and Graham, 2004). Yet other researchers have found a significant positive relationship between a lack of hardiness and burnout in women (Topf, 1989; Boyle et al., 1991; Simoni and Paterson, 1997). Once again we might conclude that while research of this nature provides us with some insight into the linkage between personality and burnout, it also raises a number of questions as to how universally applicable the findings are.

Locus of control

Locus of control has to do with an individual's beliefs about the future and who or what has the power to control future events. Individuals are said to have an internal locus of control when they believe they can control and, if necessary, change workplace events and environment. They feel that what happens to them is the result of their own actions, behavior or attitudes, and that it is largely within their power to make changes.

In contrast, individuals are said to have a locus of control that is external when they believe that forces or powers outside or beyond themselves are largely in control of their future and environment. This external power may be other people (i.e., people powerful enough to control how the future unfolds for others) or it may be thought of in terms of luck, chance or fate.

Studies have shown workers are more vulnerable to stress and burnout when they have a poorer locus of control, that is when they are more likely to assume that external forces have the power to steer the future than not (Schmitz et al., 2000). In other words, the perception of one's locus of control has been found to be a predictor for burnout (Khajeddin, Shoushtari, and Hajebi et al., 2006).

Marital status

Longer life and better mental and physical health have been linked to stable, long-term marriages. A recent study involving "one billion person years across seven European countries . . . found that married persons had age adjusted mortality rates that were 10–15 percent lower than the population as a whole" (Gallacher and Gallacher, 2011). The longer one is married, the more physical and mental health benefits accrue; men who are married have been found to fare better in terms of their physical health while women fare better than men in terms of their mental health (Gallacher and Gallacher, 2011).

On the effects of marriage on burnout, however, research has shown mixed results. Some researchers have found no link whatever between marital status and whether one is more or less likely to burn out (Gaines and Jermier, 1983; Ross et al., 1989; Gulalp et al., 2008). Yet some studies do suggest that married individuals have reported significantly lower levels of psychological burnout symptoms than their unmarried colleagues (Ifeagwazi, 2005). Like many of the socio-demographic variables discussed above, these studies often raise as many questions as they answer and call for more investigation.

Gender

Research has also sent mixed signals in terms of whether gender plays a significant role in terms of who is at risk for burnout. One study, for example, found that certain stressors, such as bad bosses, are less likely to have a detrimental effect on women's health and, interestingly, more likely to be problematic for men. When researchers looked at the hormone levels of male and female workers they found that

> women had half the levels of inflammatory proteins – the things that make you fall over clutching your chest – in their blood than men. Why are women less likely to let the bastards grind them down? It seems that when the going gets tough, the women get networking. Scientists from the University of California, Los Angeles, say that when a woman encounters stress, her body releases the hormone oxytocin, which suppresses the impulse to flee and encourages her to gather with her clan. This process releases more oxytocin, which has a calming effect. This response doesn't occur in men. They instead produce testosterone when stressed, which reduces the effects of oxytocin and triggers their fight-or-flight reflex. This only increases their stress, until they either fight, flee . . . or sit morosely in a corner and get bitterly, twistedly drunk. (Naish, 2005)

Given that other studies have found that the prevalence of burnout is likely to be much higher among women than men (e.g., Norlund et al., 2010), the above results tend only to muddy the waters further.

Purvanova and Muros (2010), for instance, have reviewed numerous burnout studies and believe that the research requires a more nuanced approach. They suggest women are more likely to report burnout and more likely to view themselves as emotionally exhausted. Men, on the other hand, are more likely to report depersonalization. When the numbers are translated into a percentage overlap statistic, 54 percent of women report experiencing emotional exhaustion compared with 46 percent of men. Likewise, 57 percent of men report experiencing depersonalization compared with only 43 percent of women.

In other words, men and women have been found to suffer from both mental and physical health problems when forced to endure chronic stress on the job. The extent to which their bodies and minds react to stress differs to some extent, however. These findings are discussed in greater depth in Chapter 7: Gender, Burnout and Work-related Stress.

Work–home interference

The specific phenomenon of one's work life having a negative impact on one's home life is called work–home interference (WHI). Work–home interference is defined as "a form of inter-role conflict in which the role pressures from the work and family domains are mutually incompatible so that participation in one role [home] is made more difficult by participation in another role [work]" (Greenhaus and Beutell, 1985, p. 77). WHI means that individuals exposed to stress each day on the job find they cannot "turn it off" once they are home.

"Interference" happens when a work deadline requires an individual to forgo a family gathering (external interference) or when one cannot stop thinking about or being preoccupied by work (internal interference). Individuals with this particular problem cannot participate fully in and enjoy home life, and find recovery from the stresses of the workday much more difficult. Studies have shown that WHI can contribute to burnout (Peeters et al., 2005) in addition to bringing about a host of adverse health problems (Frone et al., 1997).

In contrast, when matters from one's home life interfere negatively with one's work life, researchers refer to this as home–work interference (HWI). HWI affects the ability to think, concentrate and perform tasks in the workplace as required (Demerouti et al., 2007).

WHI and HWI are common phenomena. One US study found that 85 percent of all workers have ongoing family responsibilities which could potentially interfere with work (HWI), and that the proportions of men and women who report WHI problems are almost identical (Bond et al., 1998).

As with other stressors on the job, there is a toll to be paid for HWI and WHI. Concerning the latter, a 2001 National Work–Life Conflict Study conducted for Health Canada found that employees who reported work/life conflict were, in the six months before the study took place:

> 1.7 times more likely to have sought care from a mental health professional,
>
> 1.7 times more likely to have received care on an outpatient basis three or more times,
>
> 1.6 times more likely to have seen a physician three or more times,
>
> 1.5 times more likely to have visited a hospital emergency room, and
>
> 1.4 times more likely to have had to spend at least one night in the hospital. (Health Canada, 2001)

The study determined that the cost of visiting a physician as a direct result of the physical and mental stresses and health-related problems inherent in work/life conflict was approximately C$650 million per year. Inpatient hospital care brought about for the same reason amounted to approximately C$2 billion per year, and emergency department visits once again made for the same reason amounted to approximately C$144 million per year.

The study also found that employees who experienced high levels of work/life conflict were

> 5.6 times more likely to report high levels of burnout, 2.4 times more likely to report high levels of depressed mood, 2.4 times more likely to say their health is fair/poor, 2.3 times as likely to report that they were in poor physical health, and 2.2 times as likely to report high levels of perceived stress. They were slightly less than half as likely to report high levels of life satisfaction. (Health Canada, 2001)

Work/life conflict was also found to be responsible for higher rates of absenteeism and a greater desire to find new employment.

In this particular study, women were far more likely to report work/life interference; however, the link between gender and WHI has not been definitively established. Other researchers, for example, have found no difference in the rate of burnout between men and women as a result of

WHI (Putnik and Houkes, 2011) while some have found that WHI is much more strongly related to burnout in women (Peeters et al., 2005).

Interestingly, workplace conditions which could lead to work interfering with home life are far less likely to lead to burnout when an employee has a preference for those working conditions, even though they would not seem ideal to most. For example, research has shown that working long hours in and of itself is not a problem if an individual prefers working those hours (Barnett et al., 1999) and some fit exists between those hours and the individual's family life.

Expectations

There is debate about whether unrealistic expectations about a job can lead to burnout. Certain studies have concluded that higher expectations are likely to result in burnout, while others have found no connection (Kirk and Koeske, 1995; Schaufeli and Enzmann, 1998). Part of the problem with inconsistent results has to do with how expectations are defined. For example, in the literature, "expectations" have been defined to mean such disparate things as desiring positive work relationships, or greater job satisfaction, or evidence of progress in the client group served, or greater personal effectiveness, or a more challenging job, or, finally, control over one's work (Browning et al., 2006).

The idea of "expectations" can also include a search for meaning or significance in one's work. Burnout, in this context, is " 'the failure of the existential quest' – that moment when we wake up one morning and realize that what we're doing has appallingly little value" (Senior, 2006). If employees are able to convince themselves their work is meaningful, they are generally able to carry on much longer than less convinced colleagues without burning out.

However we define "expectations," it is clear that researchers see them as playing a large role in the burnout process. Alden Cass, who has worked as a therapist for high flyers on Wall Street, notes that the gulf between expectations of what a job will be like and what a job is actually like is generally responsible for the burnout he sees in clients. In that sense, the gulf between expectations and reality makes high-powered, corporate executives no different from public school teachers and others who work in less lofty positions. As Cass tells his clients, "[h]appiness is reality divided by expectations" (Senior, 2006).

Yet lowering one's expectations does not sound like a particularly desirable thing to do, either from a personal or an organizational perspective. Clearly, organizations are unlikely to want to hire candidates who have low expectations. Nonetheless, some organizations, in an attempt to deal with unrealistic expectations, have tried out or tested methods such as the so-called "realistic job preview" (RJP) and "expectation lowering procedures" (ELP) in order to "adjust" the expectations of potential and new employees.

An RJP is included as an add-on to the job interview and is designed to provide both positive and negative information about the particular position and employment situation. For example, information about the job/position can include statements such as ". . . people are often terminated for major deviations from company policy. Some people believe that the supervision is too strict" and "[s]ome common complaints about working here are the short breaks" (Buckley et al., 2002, p. 275). Studies on the efficacy of the RJP are mixed; some produce little reduction in turnover while others have been somewhat more successful (Buckley et al., 2002).

Expectation lowering procedures, in contrast, focus on an employee's expectations rather than on job-specific details. An ELP statement also accompanies a job interview and generally contains the following type of non-job specific information:

> [w]e don't want you to have "reality shock" when you discover this job isn't everything you expected it to be. When you see someone doing the job, it may not seem that bad. But once you are the one performing the work, your perspective may be quite different. The difficulties and frustrations encountered on the job are usually greater than you realize they are. You need to develop realistic expectations about this position and your opportunities with this position. (Buckley et al., 2002, p. 276)

Whether the RJP and ELP are effective tools in combating burnout is not clear, but it has been suggested that the use of either or both in combination for that purpose might be worth studying in greater detail (Halbesleben and Buckley, 2004).

Environmental and organizational factors

Workload

Workload, overload, excessive demands and time pressure, all of which have been found to contribute to burnout, are strongly related to emotional exhaustion (Schaufeli and Enzmann, 1998; Maslach et al., 2001; Bakker et al., 2003; Bakker et al., 2005). As Maslach and Leiter (1997) explain:

> [i]t is hard to find relief at work. Restful moments between events are gone. Each demand rolls without a break into the next. There is no time to catch your breath. Organizations seem to be testing the theory that people can work flat-out forever. But, with no time to recover, people soon find that their exhaustion just builds. Additional demands might be manageable if they were given more resources; extra support or equipment can turn increased demand into an opportunity. But instead, the current scramble for survival often results in a shortage of resources. (p. 39)

Overload is bad for both the individual and the organization – the quality of work drops, co-workers have no time for collegiality or to build community, and morale and motivation suffer.

Workload can be of two types – either quantitative (e.g., high job demands with too little time in which to carry out those demands) or qualitative (e.g., jobs with a great deal of complexity or requiring a great deal of concentration). Both can lead to burnout, particularly if employees have no opportunity to rest and recover between demands. Individuals who have an opportunity to rest whether at work or at home are far better able to recover from the stress of these demands (Shinn et al., 1984).

A number of models (e.g., Karasek, 1979; Hobfoll, 1989, 1998) have attempted to explain how job demands (such as workload, emotional demands and/or work pressure) can physically and psychologically affect employees and result in job strain along with numerous physiological and/or psychological consequences. Models of this sort suggest that the provision of additional resources to the employee (e.g., control, greater autonomy, opportunity for promotion, feedback and social support) can buffer the deleterious effects of workload demands and result in greater motivation and engagement.

Karasek's (1979) job strain model, for example, posits that workers who have an excessive workload are less at risk for mental and physical health issues if they are permitted greater control (i.e., the ability to decide how aspects of the work are to be done). Karasek's model has been used widely in studies on the effects of workplace stress on health.

The Job Demands–Resources Model (Bakker and Demerouti, 2007) is another model which begins with the proposition that employees are most susceptible to burnout in a work environment in which job demands are high. These demands are responsible for strain which leads to job performance problems as well as problems with health and general well-being. When employees are given additional resources (social support, the ability to vary tasks, decision-making power, feedback, rewards, job security) to meet the demands of their work, however, the outcome is greater job satisfaction, commitment and engagement in the workplace. This model, predicting that job demands play a role in burnout, and that those individuals with greater job resources are less likely to be disengaged from their work, has been supported by various studies (Bakker et al., 2005; Xanthopoulou et al., 2007).

Underwork

Underwork has been described in a number of ways including not being challenged enough on the job, or not having enough work to do, or lacking interest in the tasks that need to be done, or trying to look busy because there is not enough work. Interestingly, underwork like overwork is also a major source of workplace stress (Pines and Aronson, 1988). Certain experts who deal with the syndrome call it "boreout" – a syndrome with symptoms almost identical to burnout but which is differentiated by the fact that it afflicts the under-challenged (Uchtmann, 2012).

Peter Werder, co-author of *Diagnosis Boreout* (Rothlin and Werder, 2008), estimates that 15 per cent of all office staff are at risk of boreout/burnout because they are "seriously underchallenged." Werder and co-author Philippe Rothlin based part of their research on a study of wasted time at work carried out in 2005 by America Online and Salary.com. In this survey, 10,000 workers were polled about distractions at work. One-third of all the employees surveyed said they wasted time because they did not have enough work to do.

The study also revealed that employees were paid $759 billion per year for work which they did not do. Using the internet for personal use made up 44.7 percent of the time wasted, followed by socializing with co-workers at 23.4 percent. Workers also conducted personal business (6.8 percent), made personal phone calls (2.3 percent) or ran errands (3.1 percent) during work time.

As with burnout, individuals with boreout are often despondent, unable to sleep, lack drive and are unable to enjoy life. They also suffer from the usual physical problems associated with burnout including stomach problems, fatigue and headaches. Werder and Rothlin estimate that 15 percent of all workers are on the path to boreout (Boyes, 2007) and that fixing the situation is more difficult than one might think:

> "It is easier nowadays to confess to alcoholism than to tell your boss you are not being used properly," said co-author Philippe Rothlin. While burnout and stress are socially acceptable, boreout is seen as slacking . . . Boreout works like this: a boss refuses to delegate work, frustrated underlings ask for more to do, but are trusted only with mind-numbing tasks. Eventually they stop asking and find ways to stretch out low-intensity tasks to fill time. But mimicking work day after day erodes self-esteem. Result: the boss hurtles towards burnout while at least some of his staff edge toward boreout. The symptoms are almost identical. (Boyes, 2007)

While employees fill their time with online surfing, email and blogging, these activities do little to fill the hole left by the engagement and challenge missing from the job. Workers fear the contempt of their supervisors if they confess, yet experts believe that changing jobs might be the only other way to fix the problem when one cannot go to management.

Type of work

Burnout research originated with the study of individuals in occupations which involved "people work" – specifically where those people were needy or in trouble and required empathy (patients, inmates, students, clients, etc.) (Maslach et al., 2001). Studies have also been done of blue-collar (unskilled and skilled) and white-collar (low-medium skilled and high skilled) workers who do not work principally with people, but, instead, with things or data. Such studies have established that burnout also exists among those employee populations not employed in the human services professions (Norlund et al., 2010).

Yet the bulk of the research still takes place in the human services fields (among physicians, nurses, and teachers, for example) and, since occupation-specific burnout studies have been underway for several decades, researchers have had the opportunity to compare rates of burnout across professions. Schaufeli and Enzmann (1998), for example, compared a number of studies conducted in the United States which measured the emotional exhaustion levels of a variety of different occupations. Teachers were found to have the greatest levels of emotional exhaustion; workers in the social services and medicine had intermediate levels, and workers in mental health and post-secondary education had the lowest levels. Physicians and police officers exhibited some of the highest levels of depersonalization.

When they compared studies of the same occupational groups in the Netherlands, the authors found that the results were much the same, indicating that proneness to burnout in certain occupations is relatively predictable despite differences between countries, cultures and languages (Schaufeli and Buunk, 2002). Indeed, burnout research in countries around the world reveals far more commonalities than differences, suggesting that certain jobs along with their corresponding pressures are often the same despite geographic location.

Physical environment

The physical environment in which one works plays an important role in burnout (Demerouti et al., 2001). For example, an uncomfortable physical environment – one in which workers are exposed to noise, extremes of temperature, dirt, noxious substances, poor lighting and crowding – is one in which workers are more likely to be at risk for burnout (Pines and Aronson, 1988).

Another potential source of stress and burnout is the exposure to physical job risks such as poor air quality (Leiter, 2005) or the potential for injury or safety incidents in an unsafe workplace (Rundmo, 1996; Leiter, 2005; Day et al., 2009). Nurses who suffer higher numbers of needlestick injuries (Clarke et al., 2002), for example, or who are exposed to infected patients and/or materials (Leiter, 2005) are more likely to suffer from emotional exhaustion – one of the key elements of burnout.

But workers need not be exposed to actual risk in order to experience higher levels of stress. Even the perception of risk (i.e., the "subjective assessment of the probability of a specified type of accident happening and how concerned we are with such an event" (Marek et al., 1985, p. 152)) plays a key role in increasing stress. Accordingly, perceived risk exists alongside actual harsh physical working conditions as a predictor of the mental and physical health of employees (Eide et al., 1985; Ulleberg and Rundmo, 1997).

Cox and Tait (1991) describe a model of risk in which workers judge how significant a hazard is by assessing its lethalness (the severity of the harm that might be inflicted), prevalence (the estimation of frequency of harm), and control. A worker's estimation of the extent to which a hazard is lethal and prevalent also contributes to that worker's sense of vulnerability.

Control over environment, in contrast, allows workers to cope with workplace stress more effectively. The more workers feel they are able to control risk, the less vulnerable they are likely to feel (Ferguson et al., 1994). Proper training can also increase a worker's sense of control, as does adherence to strict safety policies and standards (Leiter and Robichaud, 1997). Not surprisingly, more control is linked to a diminished risk for burnout.

Conflict

Although conflict may occur for many reasons, the feeling that one has too much work and too few resources with which to carry it out is likely to contribute to interpersonal conflict in the workplace (Maslach et al., 2001). Such conflict is not just limited to difficulties with co-workers and supervisors; it may also involve human services recipients (patients, students, patrons, etc.) and their families.

Social support, particularly from one's peers, has been shown to help employees better endure the stresses of the workplace. It is no wonder, therefore, that conflict on the job has been shown to decrease job satisfaction (Abrahamson et al., 2009), and is associated with higher levels of burnout (Embriaco et al., 2007). Indeed, research has revealed that this kind of conflict does not have to be unrelenting to increase the risk of burnout; even periodic but emotionally intense episodes of conflict can do the same (Brotheridge and Grandey, 2002).

Role conflict and role ambiguity

Role conflict and role ambiguity have both been linked to burnout. Role ambiguity is a term used to describe the lack of clarity, certainty and/or predictability one might have expected with regards to behaviour in a job (due, perhaps to an ill-defined or ambiguous job description and/or uncertain organizational objectives). Employees who work hard on what they think is an important project only to find it shelved or placed on the back burner, as well as employees who don't understand the scope and parameters of their job, the goals they should be pursuing, and what their priorities should be are more likely to suffer from role ambiguity.

In contrast, role conflict involves competing and incompatible demands placed on an employee. This can involve the sometimes contradictory demand of being both a supervisor and a friend, the irreconcilable demands of providing good service while striving to reduce costs, or the difficulties of doing a job which is at odds with one's own values. Where role conflict and role ambiguity are high, high levels of burnout are also found (Tunc and Kutanis, 2009).

Control

An employee is said to have control when she is able to participate in and influence workplace decision-making and is also able to exercise professional autonomy. Control has to do not only with how and when the job is done, but also how much leeway employees have in determining when they can take a break or when they can schedule their vacation. Those with a significant amount of control are "much more likely to influence their workload, rewards, social interactions, and institutional justice than are people lacking control" (Leiter, 2005, p. 132).

Control in the workplace is important for both physical and mental well-being. The Whitehall Studies, led by Sir Michael Marmot, professor of epidemiology and public health at University College, London, involved a survey of 30,000 British civil servants – a group overwhelmingly middle class with free health care – over the course of three decades. The first of the Whitehall studies found that the lower the employees were in rank, the higher their rates of physical and mental illness including heart, lung, digestive and back problems, depression and suicide. Workers at the very bottom had a risk of death that was four times higher and workers in the middle had a risk of death that was two times higher than the person at the top. The main cause, according to the researchers, was ongoing stress due to a lack of autonomy in the job. Their conclusion was that giving employees more control would improve their health (Contenta, 2010).

Other studies have shown that increased job control can cut the levels of coronary heart disease, depression and absenteeism (Karasek, 1990). In addition, workers who are allowed to actively participate in workplace decision-making show higher levels of efficacy (i.e., the sense that one is performing tasks adequately and is competent at work) and lower levels of emotional exhaustion (Cherniss, 1980; Lee and Ashforth, 1990; Leiter, 1992).

In terms of burning out, workers who feel they have little job autonomy or control are at greater risk (Paine, 1982; Glass et al., 1993). Lack of control can take a number of different forms including supervisors who micromanage or organizations that do not allow employees to participate in policy or other work-related decisions (Glass et al., 1993). Lack of control is particularly stressful in an environment in which downsizing and job restructuring are taking place.

As Maslach and Leiter (1997) note, employees who lack control are generally not permitted to resolve problems as they arise, and generally must go through some straight-jacketed organizational structure on even the smallest issues. After a time, solving problems through the "proper channels" becomes more exhausting and problematic than it is worth. The organization suffers because employees start to ignore problems or to work around them, and workers suffer because of the added stress of dealing with ongoing problems.

Social support

Researchers are interested in the interaction between social support, stress management and burnout. Social support has been defined as "an exchange of resources between at least two individuals perceived by the provider or recipient to be intended to enhance the well being of the recipient" (Shumaker and Brownell, 1984). It fosters a feeling of belonging: with social support, individuals feel that they are cared for by others, can draw upon the resources of others and/or have others upon whom they can rely. This type of support can be emotional or physical; it can involve the giving of advice or the provision of tangible aid or resources.

Part of social support concerns appreciation. Pines and Aronson (1988) write,

> [t]ime and again we have found that, in a given organization, individuals hunger for appreciation. People who feel unappreciated almost never reach out to show appreciation of someone else's work. Our experience has shown that one of the best ways for individuals to encourage others to pay attention to their work is to start acknowledging the good work of others. When individuals, on their own, reach out to give each other needed support and needed appreciation, the reaching out mushrooms and grows exponentially. Moreover, peer appreciation (which is easy to institute) reduces the need for approval from above (which is frequently difficult, if not impossible, to institute). (p. 30)

Social support is said to be a major buffer against stress (Cohen and Wills, 1985; Bates and Toro, 1999) and a lack of social support has been found to be a predictor of burnout as many as ten years later (Kalimo et al., 2003).

Nonetheless, as we have noted with similar kinds of research, the results of studies on social support have not always been consistent (Prag, 2003). Some studies have found no or little connection between stress, burnout and supervisory support (Beehr and McGrath, 1992; Burke and Greenglass, 1995). Others claim to have established a link between a reported lack of supervisory support and greater ensuing risk for burnout (Seltzer and Numerof, 1988; Cheuk et al., 1994).

From the perspective of younger employees, supervisory support has also been found to be significantly associated with reducing job stress (Boyas et al., 2012). In other words, younger workers tend to rely far more on assistance and encouragement from their supervisors to help buffer higher levels of job stress. Older workers, by implication, are more self-reliant and/or have worked out other means of social support to help them cope with workplace stressors.

Research also indicates that individuals who have high levels of social support in the form of positive conversations with co-workers are less likely to experience future higher levels of emotional exhaustion and cynicism – two hallmarks of burnout (Halbesleben and Buckley, 2006). Similarly, those who have a network of colleagues with similar interests and concerns have lower scores in terms of emotional exhaustion and depersonalization (Ross et al., 1989).

Reciprocity

Reciprocity involves the amount an individual invests in a particular relationship with another and whether, and to what extent, that investment is returned. In the case of workers who feel they are contributing more to relationships with clients, patrons, patients, etc. than is being returned, there is an increased risk of burnout. For example, research involving nurses (Van Yperen et al., 1992; Schaufeli and Janczur, 1994), teachers (Van Horn et al., 2001), police officers (Kop et al., 1999), social workers (Truchot et al., 2000) and general practitioners (Bakker et al., 2000) has confirmed that there is a positive relationship between professionals who perceive a lack of reciprocity with their human services recipients and the three dimensions of burnout: emotional exhaustion, depersonalization and reduced personal accomplishment. One study of primary care physicians carried out over the course of ten years found that frequent contact with demanding patients not only leads to higher burnout scores but continues to aggravate the condition over time (Schaufeli et al., 2011).

But a perceived lack of reciprocity may exist not only with the recipients of one's care but also with colleagues. This latter situation has been found to be significantly associated with burnout (Kop et al., 1999; Van Horn et al., 2001). It is also true when such employees feel they are not benefitting as much in their relationships with their supervisors (Prins et al., 2008).

Such a perception of inequity may apply not only to co-workers and supervisors, but also to organizations. Workers naturally seek to have their contributions (time, effort, loyalty, skill) returned by their employing organization in the way of gratitude, remuneration, status, or greater input. Organizations whose workers feel a lack of this reciprocity play a key role in determining emotional exhaustion and depersonalization (Rose et al., 2010).

Where a perceived imbalance or inequity between input and output occurs, workers will attempt to correct the balance; where that balance cannot be corrected, they will experience significant distress and dissatisfaction (van Dierendonck et al., 2001). Indeed, employees who feel a lack of reciprocity with their employing organization have been found to experience an even greater distress than when that lack of reciprocity exists with patients and co-workers (Rose, 2010). As Maslach, Schaufeli and Leiter (2001) note,

> [t]he organizational context is . . . shaped by larger social, cultural, and economic forces. Recently, this has meant that organizations have undergone a lot of changes, such as downsizing and mergers, that have had significant effects on the lives of their employees. This is perhaps most evident in changes in the psychological contract – i.e. the belief in what the employer is obliged to provide based on perceived promises of reciprocal exchange (Rousseau, 1995). Now employees are expected to give more in terms of time, effort, skills, and flexibility, whereas they receive less in terms of career opportunities, lifetime employment, job security, and so on. Violation of the psychological contract is likely to produce burnout because it erodes the notion of reciprocity, which is crucial in maintaining well-being. (p. 409)

In order to be viewed as a satisfying relationship, the investments and outcomes must be perceived by the employee as being more or less fair and equal.

Social comparison

Researchers have found a link between social comparison and burnout. Social comparison involves the comparison employees make between themselves and their peers. Employees will, in theory, experience stress when they see that some of their peers are paid more or appear to be better treated than themselves.

Social comparison research involves such equations as: "when I see colleagues who experience more difficulties than I do, I am happy that I am doing so well myself;" "when I see colleagues who are doing worse [than I am], I experience fear that my future will be similar;" "when I see colleagues who are doing better than I am, I feel frustrated about my own situation;" and "when I see colleagues who are doing better than I am, I realize that it is possible to improve" (Carmona et al., 2006, p. 90). Research has established that employees who identify with those who appear to be doing worse than themselves and contrast themselves with those who are doing better are more likely to be at risk of burnout (Brenninkmeijer et al., 2000; Carmona et al., 2006).

Leadership style

It is, of course, no surprise to learn that an employee's supervisor/manager can be a source of negative emotions in the workplace. There is an established link between type of leadership style, negative emotions, stress and burnout. One study involving employees reporting negative and positive emotions at work found that 7 percent of positive emotional events were attributable to "acts of management," as were 22 percent of negative emotional events (Basch and Fisher, 2000). Employees who were asked to remember an emotional event in the workplace involving their supervisor/manager were more likely to remember one which involved negative rather than positive emotions (Dasborough, 2006).

Leadership style in terms of credibility has to do with the extent to which workers feel they can trust the person in charge. The extent to which a leader is considered credible or trustworthy involves such factors as how openly leaders communicate with their employees, whether they do what they've promised to do, and whether they explain what they find important and what their vision is (Gabris and Ihrke, 1996, p. 233; Kouzes and Posner, 1987, 1993). Where workers feel they can trust their managers/supervisors, they are less likely to question managerial objectives or to refrain from volunteering to take on additional work required to accomplish these objectives. Where leadership credibility is low, the extent to which employees are burned out is likely to be higher (Gabris and Ihrke, 1996).

Researchers have sought to categorize and identify a leadership style which is less likely to result in employee stress. Bass (1998), for example, describes three types of leadership style: transformational, transactional, and laissez-faire. A transformational leadership style is one in which the leader attempts to inspire and motivate by appealing to an individual's sense of self-worth in addition to other unique emotional and developmental needs. Such a leader may use material as well as psychological rewards, but the emphasis is on the psychological and includes positive feedback and praise.

In contrast, transactional leadership, according to Bass, is of three types. These are, in his terminology: contingent reward, management by exception-active, and management by exception-passive. "Contingent reward" involves the provision of a reward (money, positive feedback, commendation) in exchange for good job performance. "Management by exception" may be active or passive. In the active form, the leader monitors an employee's work, and if job performance deviates from expected performance standards, correction in the shape of negative feedback, discipline or punishment takes place. In the passive form, the leader is not actively involved in the monitoring of work, and only steps in when any deviation is brought to his or her attention. Laissez-faire leadership is, in contrast to the transformational and transactional models described above, the absence of leadership.

Research has linked this matter of leadership style to the perception of job-related stress in employees (Sosik and Godshalk, 2000). Specifically, transformational leadership has been associated with enhanced task performance and the sense on the part of employees that they had greater social support than perceived with other leadership styles. In addition, employees working with a transformational leader rather than a transactional leader are less likely to see an impending stressor as threatening. Transformational leadership has been found to increase well-being and job satisfaction, and to decrease burnout in employees (Weberg, 2010).

Conclusion

A number of links have been established between various individual, socio-demographic and workplace characteristics and the potential risk for burnout. Some of these links are tenuous; others are clear cut and well established. An excessive workload, for example, has been shown to have a significant connection with burnout as has having little control over how the job is done. Other factors, such as lack of social support from colleagues or superiors, have been shown in some cases to be a important predictors of the syndrome.

For Leiter and Maslach (2005), burnout is more than merely the alignment of risk factors, but can instead be explained by the level of "mismatch" between an individual and six strategic areas of the workplace: workload, control, reward, community, fairness and values. These authors suggest that many of the factors discussed above fall into one of these six categories. Even certain individual and socio-demographic factors could be said to belong, to some extent, to some of these categories.

In Chapter 8, Symptom Recognition and Preventing Burnout, we look specifically at these six strategic areas and discuss how to handle, cope with or eliminate the various factors that contribute to burnout. In our next chapter, however, we turn to information professionals specifically and how budget cuts, downsizing and other pressures have added to the stresses of the workplace.

References

Abrahamson, K., Suitor, J., and Pillemer, K. (2009) Conflict between nursing home staff and residents. *Journal of Aging and Health*, 21(6), 895–912.

Ahola, K., Honkonen, T., Virtanen, M., Aromaa, A., and Lönnqvist, J. (2008) Burnout in relation to age in the adult working population. *Journal of Occupational Health*, 50(4), 362–5.

Bakker, A.B. and Demerouti, E. (2007) The job demands–resources model: State of the art. *Journal of Managerial Psychology*, 22, 309–28.

Bakker, A.B., Demerouti, E., and Schaufeli, W.B. (2003) Dual processes at work in a call centre: An application of the job demands–resources model. *European Journal of Work and Organizational Psychology*, 12(4), 393–417.

Bakker, A.B., Demerouti, E., and Euwema, M.C. (2005) Job resources buffer the impact of job demands on burnout. *Journal of Occupational Health Psychology*, 10, 170–80.

Bakker, A.B., Schaufeli, W.B., Sixma, H.J., Bosveld, W., and Dierendonck, D. (2000) Patient demands, lack of reciprocity, and burnout: A five-year longitudinal study among general practitioners. *Journal of Organizational Behavior*, 21(4), 425–41.

Barber, C.E. and Iwai, M. (1996) Role conflict and role ambiguity as predictors of burnout among staff caring for elderly dementia patients. *Journal of Gerontological Social Work*, 26, 101–16.

Barnett, R.C., Gareis, K.C., and Brennan, R.T. (1999) Fit as a mediator of the relationship between work hours and burnout. *Journal of Occupational Health Psychology*, 4, 307–17.

Basch, J. and Fisher, C.D. (2000) Affective events–emotion matrix: A classification of work events and associated emotions. In N.M. Ashkanasy, C.E.J. Härtel, and W.J. Zerbe (Eds.), *Emotions in the Workplace: Research, Theory, and Practice* (pp. 36–48). Westport, Connecticut: Quorum.

Bass, B.M. (1998) *Transformational Leadership: Industrial, Military, and Educational Impact*. Mahwah, New Jersey: Erlbaum.

Bates, D.S., and Toro, P.A. (1999) Developing measures to assess social support among homeless and poor people. *Journal of Community Psychology*, 27, 137–56.

Beehr, T.A., and McGrath, J.E. (1994) Social support, occupational stress and anxiety. *Anxiety, Stress and Coping*, 5, 7–19.

Bennett, L. and Michie, P. (1991) Quantitative analysis of burnout and its associated factors in AIDS nursing. *AIDS Care*, 3(2), 181–93.

Bond, J.T., Galinsky, E., and Swanberg, J.E. (1998) *The 1997 National Study of the Changing Workforce*. New York, New York: Families and Work Institute.

Boyas, J., Wind, L.H., and Kang, S. (2012) Exploring the relationship between employment-based social capital, job stress, burnout, and intent to leave among child protection workers: An age-based path analysis model. *Children and Youth Services Review*, 34, 50–62.

Boyes, R. (15 September 2007) "Boreout" emerging as new office disease; An estimated 15 percent of workers on path to ailment. *Calgary Herald*, A22.

Boyle, A., Grap, M., Younger, J., and Thornby, D. (1991) Personality hardiness, ways of coping, social support and burnout in critical care nurses. *Journal of Advanced Nursing*, 16, 850–7.

Brenninkmeyer, V., Van Yperen, N., and Buunk, B.P. (2000) I am not a better teacher, but others are doing worse: Burnout and perceptions of superiority among teachers. *Social Psychology of Education*, 4(3–4), 259–74.

Brewer, E.W. and Shapard, L. (2004) Employee burnout: A meta-analysis of the relationship between age or years of experience. *Journal of Human Resource Development Review*, 3(2), 102–23.

Brotheridge, C.M. and Grandey, A.A. (2002) Emotional labor and burnout: Comparing two perspectives of people work. *Journal of Vocational Behavior*, 60, 17–39.

Browning, L., Ryan, C., Greenberg, M., and Rolniak, S. (2006) Effects of cognitive adaptation on the expectation–burnout relationship among nurses. *Journal of Behavioral Medicine*, 29(2), 139–50.

Buckley, M.R., Mobbs, T.A., Mendoza, J.L., Novicevic, M.M., Carraher, S.M., and Beu, D.S. (2002) Implementing realistic job previews and expectation-lowering procedures: A field experiment. *Journal of Vocational Behavior*, 61, 263–78.

Burisch, M. (2006) *Das Burnout-Syndrom: Theorie der inneren Erschöpfung* [The burnout-syndrome: A theory of inner exhaustion]. Heidelberg, Germany: Springer Medizin Verlag.

Burke, R.J. and Greenglass, E. (1995) A longitudinal study of psychological burnout in teachers. *Human Relations*, 48(2), 187–202.

Burke, R.J., and Richardson, A.M. (1996) Stress, burnout and health. In C.L. Cooper (Ed.), *Stress, Medicine and Health* (pp. 101–17). Boca Raton, Florida: CRC Press.

Cano-García, F.J., Padilla-Muñoz, E.M., and Carrasco-Ortiz, M.A. (2005) Personality and contextual variables in teacher burnout. *Personality and Individual Differences*, 38(4), 929–40.

Carmona, C., Buunk, A.P., Peiró, J.M., Rodríguez, I., and Bravo, M.J. (2006) Do social comparison and coping styles play a role in the development of burnout? Cross-sectional and longitudinal findings. *Journal of Occupational and Organizational Psychology*, 79, 85–99.

Casserley, T.B., and Megginson, D. (2009) *Learning from Burnout: Developing Sustainable Leaders and Avoiding Career Derailment.* Oxford, UK: Butterworth-Heinemann.

Cherniss, C. (1980) *Professional Burnout in Human Service Organizations.* New York, New York: Praeger.

Cheuk, W.H., Wong, K.S., and Rosen, S. (1994) The effects of spurning and social support on teacher burnout. *Journal of Social Behavior and Personality*, 9, 657–64.

Clarke, S.P., Sloane, D.M., and Aiken, L.H. (2002) Effects of hospital staffing and organizational climate on needlestick injuries to nurses. *American Journal of Public Health*, 92(7), 1115–19.

Cohen, S., and Wills, T.A. (1985) Stress, social support, and the buffering hypothesis. *Psychological Bulletin*, 98, 310–57.

Contenta, S. (23 January 2010) "21st century syndrome" or just a case of the blahs? An intriguing diagnosis posits that we live in an age so stressful we can no longer cope. *Toronto Star*, IN01.

Cordes, C.L., and Dougherty, T.W. (1993) A review and an integration of research on job burnout. *Academy of Management Review*, 18, 621–56.

Cox, S.J., and Tait, N.P. (1991) *Reliability, Safety and Risk Management. An Integrated Approach.* London, UK: Butterworth-Heinemann.

Dasborough, M.T. (2006) Cognitive asymmetry in employee emotional reactions to leadership behaviors. *The Leadership Quarterly*, 17, 163–78.

Day, A.L., Sibley, A., Scott, N., Tallon, J.M., and Ackroyd-Stolarz, S. (2009) Workplace risks and stressors as predictors of burnout: The moderating impact of job control and team efficacy. *Canadian Journal of Administrative Sciences*, 26(1), 7–22.

Demerouti, E., Bakker, A.B., Nachreiner, F., and Schaufeli, W.B. (2001) The job demands–resources model of burnout. *Journal of Applied Psychology*, 86(3), 499–512.

Demerouti, E., Taris, T.W., and Bakker, A.B. (2007) Need for recovery, home–work interference and performance: Is lack of concentration the link? *Journal of Vocational Behavior*, 71, 204–20.

Dillon, J.F., and Tanner, G.R. (1995) Dimensions of career burnout among educators. *Journalism & Mass Communication Educator*, 50(2), 4–13.

D'Souza, F., Egan, S.J., and Rees, C.S. (2011) The relationship between perfectionism, stress and burnout in clinical psychologists. *Behaviour Change*, 28(1), 17–27.

Duffy, B., Oyebode, J.R., and Allen, J. (2009) Burnout among care staff for older adults with dementia: The role of reciprocity, self-efficacy and organizational factors. *Dementia*, 8, 515–41.

Eide, R., Hellesøy, O.H., and Gogstada, C. (1985) Associations between experiences of environmental factors and health. In O.H. Hellesøy (Ed.), *Work Environment Statfjord Field: Work Environment, Health, and Safety on a North Sea Oil Platform* (pp. 317–24). Oslo, Norway: Universitetsforlaget.

Embriaco, N., Azoulay, E., Barrau, L., Kentish, N., Pochard, F., Loundou, A., and Papazian, L. (2007) High level of burnout in intensivists: Prevalence and associated factors. *American Journal of Respiratory and Critical Care Medicine*, 175(7), 686–92.

Fenlason, K.J., and Beehr, T.A. (1994) Social support and occupational stress: Effects of talking to others. *Journal of Organizational Behavior*, 15(2), 157–175.

Ferguson, E., Cox, T., Farnsworth, W., Leiter, M., and Irving, K. (1994) Nurses' anxieties about biohazards as a function of context and knowledge. *Journal of Applied Social Psychology*, 24(10), 926–40.

Fontana, D. and Abouserie, R. (1993) Stress levels, gender and personality factors in teachers. *British Journal of Educational Psychology*, 63, 261–70.

Frone, M.R., Russell, M., and Cooper, M.L. (1997) Relation of work–family conflict to health outcomes: A four-year longitudinal study of employed parents. *Journal of Occupational and Organizational Psychology*, 70, 325–35.

Gabris, G.T., and Ihrke, D.M. (1996) Job burnout in a large federal agency: Phase model implications for how employees perceive leadership credibility. *Public Administration Quarterly*, 11(2), 220–49.

Gaines, J., and Jermier, J.M. (1983) Emotional exhaustion in a high stress organization. *Academy of Management Journal*, 26(4), 567–86.

Gallacher, D., and Gallacher, J. (2011) Are relationships good for your health? *Student BMJ*, 19:d404. Retrieved from *Student BMJ*: *http://student.bmj.com/student/view-article.html?id=sbmj.d404.*

Ghorpade, J., Lackritz, J., and Singh, G. (2011) Personality as a moderator of the relationship between role conflict, role ambiguity, and burnout. *Journal of Applied Social Psychology*, 41, 1275–98.

Glass, D.C., McKnight, J.D., and Valdimarsdottir, H. (1993) Depression, burnout, and perceptions of control in hospital nurses. *Journal of Consulting and Clinical Psychology*, 61(1), 147–55.

Glouberman, D. (2002) *The Joy of Burnout.* Isle of Wight: Skyros Books.

Greenhaus, J.H. and Beutell, N.J. (1985) Sources of conflict between work and family roles. *Academy of Management Review*, 10, 76–88.

Gulalp, B., Karcioglu, O., Sari, A., and Koseoglu, Z. (2008) Burnout: Need help? *Journal of Occupational Medicine and Toxicology*, 3(1), 32.

Haddad, A. (1998) Sources of social support among school counselors in Jordan and its relationship to burnout. *International Journal for the Advancement of Counseling*, 20, 113–21.

Halbesleben, J.R.B., and Buckley, M.R. (2004) Burnout in organizational life. *Journal of Management*, 30(6), 859–79.

Halbesleben, J., and Buckley, M.R. (2006) Social comparison and burnout: The role of relative burnout and received social support. *Anxiety, Stress & Coping*, 19(3), 259–78.

Health Canada (2001) Work–life conflict in Canada in the new millennium: Executives Summary. Retrieved from *http://www.hc-sc.gc.ca/ewh-semt/pubs/occup-travail/balancing_six-equilibre_six/sum-res-eng.php.*

Hobfoll, S.E. (1998) *Stress, Culture, and Community.* New York, New York: Plenum Press.

Hobfoll, S.E. (1989) Conservation of resources: A new attempt at conceptualizing stress. *American Psychologist*, 44(3), 513–24.

Ifeagwazi, C.M. (2005) The influence of marital status on self-report of symptoms of psychological burnout among nurses. *Omega*, 52(4), 359.

Jackson, R.A. (1993) An analysis of burnout among school of pharmacy faculty. *American Journal of Pharmaceutical Education*, 57(1), 9–17.

Kalimo, R., Pahkin, K., Mutanen, P., and Toppinen-Tanner, S. (2003) Staying well or burning out at work: Work characteristics and personal resources as long-term predictors. *Work & Stress*, 17, 109–22.

Kantas, A., and Vassilaki, E. (1997) Burnout in Greek teachers: Main findings and validity of the Maslach Burnout Inventory. *Work & Stress*, 11(1), 94–100.

Karasek, R.A. (1979) Job demands, job decision latitude, and mental strain: Implications for job redesign. *Administrative Science Quarterly*, 24, 285–308.

Karasek, R. (1990) Lower health risk with increased job control among white collar workers. *Journal of Organizational Behavior*, 11(3), 171–85.

Khajeddin, N., Shoushtari, M., and Hajebi, A. (2006) The impact of perception of locus of control on burnout syndrome among nurses in a psychiatric hospital. *Iranian Journal of Psychiatry and Clinical Psychology*, 12(1), 43–8.

Kirk, S.A., and Koeske, G.F. (1995) The fate of optimism – A longitudinal study of case managers' hopefulness and subsequent morale. *Research on Social Work Practice*, 5(1), 47–61.

Klag, S., and Graham, B. (2004) The role of hardiness in stress and illness: An exploration of the effect of negative affectivity and gender. *British Journal of Health Psychology*, 9, 137–61.

Konert, E. (1997) *Relationship among middle-school teacher burnout, stress, job satisfaction and coping styles.* Unpublished PhD Dissertation. Detroit, Michigan: Wayne State University.

Kop, N., Euwema, M., and Schaufeli, W. (1999) Burnout, job stress and violent behavior among Dutch police. *Work & Stress*, 13, 326–40.

Kouzes, J.M., and Posner, B.Z. (1987) *The Leadership Challenge: How to Get Extraordinary Things Done in Organizations.* San Francisco, California: Jossey-Bass.

Kouzes, J.M., and Posner, B.Z. (1993) *Credibility: How Leaders Gain and Lose it, Why People Demand it.* San Francisco, California: Jossey-Bass.

Lee, R.T., and Ashforth, B.E. (1990) On the meaning of Maslach's three dimensions of burnout. *Journal of Applied Psychology*, 75, 743–7.

Leiter, M.P. (1992) Burn-out as a crisis in self-efficacy: Conceptual and practical implications. *Work & Stress*, 6(2), 107–15.

Leiter, M.P. (2005) Perception of risk: An organizational model of burnout, stress symptoms, and occupational risk. *Anxiety, Stress, and Coping*, 18, 131–44.

Leiter, M.P., and Robichaud, L. (1997) Relationships of occupational hazards with burnout: An assessment of measures and models. *Journal of Occupational Health Psychology*, 2(1), 35–44.

Lyons, J.B., and Schneider, T.R. (2009) The effects of leadership style on stress outcomes. *The Leadership Quarterly*, 20(5), 737.

Marek, J., Tangernes, B., and Hellesøy, O.H. (1985) Experience of risk and safety. In O.H. Hellesøy (Ed.), *Work Environment Statfjord Field: Work Environment, Health, and Safety on a North Sea Oil Platform*. Oslo, Norway: Universitetsforlaget.

Maslach, C., and Leiter, M.P. (1997) *The Truth about Burnout: How Organizations Cause Personal Stress and What to Do about it*. San Francisco, California: Jossey-Bass.

Maslach, C., Schaufeli, W.B., and Leiter, M.P. (2001) Job burnout. In S.T. Fiske, D.L. Schachter, and C. Zahn-Waxler (Eds.), *Annual Review of Psychology*, 52, 397–422.

Mills, L.B. and Huebner, E.S. (1998) A prospective study of personality characteristics, occupational stressors, and burnout among school psychology practitioners. *Journal of School Psychology*, 36, 103–20.

Naish, J. (24 November 2005) Killer bosses. *The Times*, Career 4.

Norlund, S., Reuterwall, C., Höög, J., Lindahl, B., Janlert, U., and Birgander, L.S. (2010) Burnout, working conditions and gender – results from the northern Sweden MONICA Study. *BMC Public Health*, 10, 326.

Paine, W.S. (1982) *Job Stress and Burnout: Research, Theory, and Intervention Perspectives*. Beverly Hills, California: Sage.

Peeters, M.C.W., Montgomery, A.J., Bakker, A.B., and Schaufeli, W.B. (2005) Balancing work and home: How job and home demands are related to burnout. *International Journal of Stress Management*, 12(1), 43–61.

Pines, A., and Aronson, E. (1988) *Career Burnout: Causes and Cures*. New York, New York: The Free Press.

Prag, P.W. (2003) Stress, burnout, and social support: A review and call for research. *Air Medical Journal*, 22(5), 18–22.

Prins, J.T., Gazendam-Donofrio, S.M., Dillingh, G.S., van de Wiel, H.B.M., van der Heijden, F.M.M.A., and Hoekstra-Weebers, J.E.H.M. (2008) The relationship between reciprocity and burnout in Dutch medical residents. *Medical Education*, 42(7), 721–8.

Purvanova, R.K., and Muros, J.P. (2010) Gender differences in burnout: A meta-analysis. *Journal of Vocational Behavior*, 77(2), 168–85.

Putnik, K., and Houkes, I. (2011) Work related characteristics, work–home and home–work interference and burnout among primary healthcare physicians: A gender perspective in a Serbian context. *BMC Public Health*, 11, 716. Retrieved from *http://www.biomedcentral.com/1471-2458/11/716*.

Ray, E.B., and Miller, K.I. (1994) Social support, home/work stress, and burnout: Who can help? *Journal of Applied Behavioral Science*, 30(3), 357–73.

Roberts, G.A. (1997) Prevention of burn-out. *Advances in Psychiatric Treatment*, 3, 282–9.

Rose, J., Madurai, T., Thomas, K., Duffy, B., and Oyebode, J. (2010) Reciprocity and burnout in direct care staff. *Clinical Psychology & Psychotherapy*, 17(6), 455–62

Ross, R., Altmaier, E., and Russell, D. (1989) Job stress, social support, and burnout among counseling centre staff. *Journal of Counseling Psychology*, 36(4), 464–70.
Rothlin, P., and Werder, P. (2008) *Boreout: Overcoming Workplace Demotivation*. London, UK: Kogan Page.
Rousseau, D. (1995) *Psychological Contracts in Organizations: Understanding Written and Unwritten Agreements*. Thousand Oaks, California: Sage.
Rundmo, T. (1996) Associations between risk perception and safety. *Safety Science*, 24(3), 197–209.
Schaufeli, W.B., and Buunk, B.P. (2002) Burnout: An overview of 25 years of research and theorizing. In M.J. Schabracq, J.A.M. Winnubst, and C.L. Cooper (Eds.), *The Handbook of Work and Health Psychology*, 2nd ed. Chichester, UK: Wiley.
Schaufeli, W.B., and Enzmann, D. (1998) *The Burnout Companion to Study and Practice: A Critical Analysis*. London, UK: Taylor and Francis.
Schaufeli, W.B., and Janczur, B. (1994) Burnout among nurses: A Polish–Dutch comparison. *Journal of Cross-Cultural Psychology*, 25, 95–113.
Schaufeli, W.B., Maassen, G.H., Bakker, A.B., and Sixma, H.J. (2011) Stability and change in burnout: A 10-year follow-up study among primary care physicians. *Journal of Occupational and Organizational Psychology*, 84(2), 248.
Schmitz, N., Neumann, W., and Oppermann, R. (2000) Stress, burnout and locus of control in German nurses. *International Journal of Nursing Studies*, 37(2), 95–9.
Schwartz, R.H., Tiamiyu, M.F., and Dwyer, D.J. (2007) Social worker hope and perceived burnout. *Administration in Social Work*, 31(4), 103–19.
Seltzer, J., and Numerof, R.E. (1988) Supervisory leadership and subordinate burnout. *Academy of Management Journal*, 31, 439–46.
Senior, J. (26 November 2006) Can't get no satisfaction: In a culture where work can be a religion, burnout is its crisis of faith. Retrieved from *http://nymag.com/news/features/24757/*.
Shinn, M., Rosario, M., Morch, H., and Chestnut, D.E. (1984) Coping with job stress and burnout in the human services. *Journal of Personality and Social Psychology*, 46, 864–76.
Shumaker, S.A., and Brownell, A. (1984) Toward a theory of social support: Closing conceptual gaps. *Journal of Social Issues*, 40(4), 11–36.
Simoni, P., and Paterson, J. (1997) Hardiness, coping, and burnout in the nursing workplace. *Journal of Professional Nursing*, 13(3), 178–85.
Sosik, J.J., and Godshalk, V.M. (2000) Leadership styles, mentoring functions received, and job-related stress: A conceptual model and preliminary study. *Journal of Organizational Behavior*, 21, 365–90.
Thomsen, S., Soares, J., Nolan, P., Dallender, J., and Arnetz, B. (1999) Feelings of professional fulfillment and exhaustion in mental health personnel. *Psychotherapy and Psychosomatics*, 68(3), 157–64.
Topf, M. (1989) Personality, hardiness, occupational stress, and burnout in critical care nurses. *Research in Nursing & Health*, 12, 179–88.
Truchot, D., Keirseblick, L., and Meyer, S. (2000) Communal orientation may not buffer burnout. *Psychological Reports*, 86, 872–8.

Tunc, T., and Kutanis, R.O. (2009) Role conflict, role ambiguity, and burnout in nurses and physicians at a university hospital in Turkey. *Nursing & Health Sciences*, 11(4), 410–16.

Uchtmann, J. (22 January 2012) Underchallenged and underworked. *The Nation*.

Ulleberg, P., and Rundmo, T. (1997) Job stress, social support, job satisfaction and absenteeism among offshore oil personnel. *Work & Stress*, 11(3), 215–28.

van Dierendonck, D., Schaufeli, W.B., and Buunk, B.P. (2001) Burnout and inequity among human service professionals: A longitudinal study. *Journal of Occupational Health Psychology*, 6(1), 43–52.

Van Horn, J.E., Schaufeli, W.B., and Taris, T.W. (2001) Lack of reciprocity among Dutch teachers: Validation of reciprocity indices and their relation to stress and wellbeing. *Work & Stress*, 15(3), 191–213.

Van Yperen, N.W., Buunk, A.P., and Schaufeli, W.B. (1992) Communal orientation and the burnout syndrome among nurses. *Journal of Applied Social Psychology*, 22, 173–89.

Watson, D., Clark, L.A., and Harkness, A.R. (1994) Structures of personality and their relevance to psychopathology. *Journal of Abnormal Psychology*, 103, 18–31.

Weberg, D. (2010) Transformational leadership and staff retention: An evidence review with implications for healthcare systems. *Nursing Administration Quarterly*, 34(3), 246–58.

Williams, C.A. (1989) Empathy and burnout in male and female helping professionals. *Research in Nursing & Health*, 12, 169–78.

Xanthopoulou, D., Bakker, A.B., Demerouti, E., and Schaufeli, W.B. (2007) The role of personal resources in the job demands–resources model. *International Journal of Stress Management*, 14(2), 121–41.

Yildirim, I. (2008) Relationships between burnout, sources of social support and sociodemographic variables. *Social Behavior and Personality*, 36(5), 603–16.

Zellars, K.L., Perrewé, P.L., and Hochwarter, W.A. (2000) Burnout in health care: The role of the five factors of personality. *Journal of Applied Social Psychology*, 30, 1570–98.

3

Burnout and information professionals

How we got this way

Abstract: This chapter analyses the particular sources of workplace stress for information professionals. These sources include the profession's poor image and status, as well as the role gender plays. Consideration is given to the problems of poor management, shrinking budgets, downsizing, and the shift from print to electronic sources as potential sources of stress. Finally, the increase in attacks, assaults and low-level nuisances are discussed.

Key words: burnout, librarianship, image, status, gender, management, budgets, downsizing, print and electronic sources, assaults and nuisances.

Introduction

It would be startling to come across an article predicting the end of policing, nursing, teaching or firefighting as professions. That's not true, unfortunately, for librarians. On a regular basis, those who work in the information professions see articles predicting the end of libraries, librarians and books themselves. Even at symposia aimed specifically at and attended mainly by beleaguered librarians, invited speakers frequently declare key areas of librarianship "dead," and recommend hiring computer people instead.

A case in point: at a recent Connecticut Library Association symposium, for example, Eli Neiburger, Associate Director for IT and Production at Ann Arbor District Library, told the audience that libraries no longer needed reference librarians. Library patrons, he said, were now accustomed to doing their own searches online for information, and reference librarians, like travel agents, were a thing of the past. Just as individuals could book their own trips online, they could also find their own answers online – reference was, he declared, dead. Instead, libraries needed to take greater control of their digital infrastructure and resources which required, in essence, more computer experts to do the job (Kelley, 2011).

Not only is reference dead, but so are public libraries according to Marc Brodnick in a *Forbes* piece. For Brodnick, public libraries are on the path to extinction because they exist, primarily, to lend books. When enough people own tablet devices and e-readers, he asserts, they will simply download books (i.e., purchase them cheaply) and stop using libraries. "I believe strongly," Brodnick writes, "that public libraries will turn into ghost towns (sic) in five to fifteen years, at which point it will become very difficult to justify funding them and keeping them open" (Brodnick, 2012).

In a recent Salon.com article, Martha Nichols wonders if libraries are dead but for different reasons. Nichols is concerned that libraries may have reached the end of their usefulness, not because books are available online, but because they have lost their way. She mourns what they have become in their efforts to "create a 'hybrid' that would mix the qualities of a library and a retail bookstore" (Nichols, 2010). The overflowing stacks of the traditional library once attracted readers. New libraries, on the other hand, with their architecturally designed empty spaces along with coffee shops, meeting spaces, and rows of computer banks, appear designed to attract the laptop user rather than the book reader. For her, books in these new libraries are used as a "design element, like potted plants" (Nichols, 2010) rather than existing as the central focus of the enterprise. Traditional libraries once stood for something important, Nichols notes – "a vision of information that's not constantly threatening to overwhelm us." Nichols and others (e.g., Lloyd, 2010) believe that modern libraries have lost that vision: the latest incarnation of the library has little to recommend it and, instead, requires us to ask what its purpose is.

Such predictions about the end of librarians and their libraries pale in comparison, however, to predictions about the end of books. At the Edinburgh International Book Festival, for example, author Ewan Morrison (2011) predicted the end of paper books within 25 years. Granted, such dire predictions have been around for some time, according to Price (2012) in the *New York Times*. Indeed, a subset of every generation seems certain that the end of print books is near.

Prognostications concerning the end of libraries aside, it's no wonder information professionals are stressed. They have had to contend with ongoing budget cuts, staff cuts, crumbling infrastructure, problems with image and status, and with the shift from print to digital collections. Perhaps most alarming of all has been the rise of violence in libraries open to the public, the risk of which has added significantly to an already onerous environment.

Ironically, information professionals also have to contend with the popular notion that library work is one of the least stressful jobs in the workplace. A 2010 ranking of the ten least stressful jobs rated librarians as number seven, only slightly more stressful than musical instrument repairers, medical records technicians, actuaries, forklift operators, appliance repairmen and medical secretaries; but they were less hassled than bookkeepers, piano tuners and janitors (CareerCast, 2010). Similar sentiments abound elsewhere: an article on low stress jobs in the *Sydney Morning Herald* for example noted:

> [t]he perception that being a librarian is a breeze appears to be grounded in truth. In a library's serene sphere, you are not constantly forced to be accountable and comply with social norms, so your anxiety and stress levels drop, organisational psychologist Christopher Shen says. A piqued reaction to a fine may be as harried as it gets. (Wilson, 2009)

Such misinformed and patronising comments are irritating to librarians. But even if they grant that in former times their profession may have seemed "a breeze," it's hard to believe any rational outside observer today would fail to notice the gale force winds that are now blowing. We will consider here the problems of budgets, the shift from print to electronic information, and, most worrisome, the increase in violence in libraries. We begin, though, with the perennial problem of the profession's image and status.

Sources of workplace stress

Image, status, and gender

In 1992, a survey of librarians in countries around the world revealed that 82 percent of librarians agreed or strongly agreed with the statement

"The status of the library profession is low" (Prins and de Gier, 1992, p. 111). Respondents explained that their profession was largely invisible, i.e., that the general public knew (and cared) little about the type of work that librarians did. They also felt that fellow employees in the same organizations (schools, government, businesses, etc.), in addition to the heads of those organizations knew little or nothing about the jobs of librarians. This lack of knowledge meant those in positions of authority placed little value on their contribution; this, in turn, was responsible for low salaries. Lack of knowledge about what librarians did also left libraries at greater risk of having their budgets slashed and their staffs downsized when those holding the purse strings needed to trim costs.

Respondents to this survey also thought that librarianship was a low status profession because it was neither dynamic nor essential to a country's *economic* growth. Particularly in developing nations, respondents worried that those outside the profession saw little or no connection between libraries and the growth or improvement of national economies. Instead, libraries were more likely to be viewed as a frivolity or a drain on struggling or emerging nations' resources (Prins and de Gier, 1992).

Survey participants believed that most people could not distinguish between librarians, library volunteers (i.e., people who worked for free) and other non-professional library staff. There was also the problem that librarians' jobs were often vastly different from one another. For example, someone employed as the head of a national library would perform none of the same tasks as someone who worked in a cataloguing department, yet both would call themselves "librarians" (Prins and de Gier, 1992). Given such a vast divergence in their respective job descriptions, the researchers responsible for this survey wondered if librarianship could truly call itself a "profession."

They noted that the kinds of work we usually think of as true professions – e.g. medicine, law and engineering – required rigorous training of their initiates which resulted in a set of skills that could not be replicated by members of the general public. The work these professionals performed was considered important enough to warrant specific legislation in most countries to prevent unscrupulous individuals from passing themselves off as members of these professions (Prins and de Gier, 1992).

The work of a librarian, in contrast, can as often as not be performed by individuals who have not gone to library school. Librarians, unlike physicians who often only do the most skilled parts of the job and leave less skilled tasks to supporting medical staff, do a number of tasks each day which require little skill or education. Again, the designation "librarian" is not one protected by legislation as it is with other professions. As a result, anyone can legitimately declare that he or she is a librarian without ever having been specifically trained for the job (Prins and de Gier, 1992).

Librarianship, then, was seen as lacking many of the hallmarks of other "true" professions. Survey participants worried that the training for librarianship as a profession did not appear as rigorous as other professions and did not attract the same caliber of students. Indeed, respondents described library science as a discipline which attracted "second or third rate students for whom the LIS represents the last chance to become graduate students" (Prins and de Gier, 1992, p. 117). Library and information science schools were also thought by survey participants to be staffed by second rate instructors who subscribed to an outdated and obsolete curriculum.

Finally, the fact that librarianship was still largely a woman's occupation also contributed to its low status, according to those who answered the survey. Sadly, in both industrialized and developing nations around the globe, jobs in which women made up the majority of the workforce were generally viewed as inferior, and librarianship was no different. In addition, because library work had been traditionally viewed as clerical, it was felt in many parts of the globe to be more suited to women (Golub, 2009).

One might logically think, then, that since it is a profession which largely employs women, at least women would rise to the top and work in those senior management roles traditionally staffed by men. Ironically, that is not at all the case. From the moment men began to enter the library profession, they disproportionately shot to the top management positions. In 1983, for example, 78.3 percent of the members of the American Library Association were female and 21.7 percent were male (Heim and Estabrook, 1983; Golub, 2009). Despite these percentages, 49.6 percent of men were in administrative positions compared with 30.4 percent of women. More alarming, however, was that men consistently beat out women in terms of salary "even when personal, professional, and organizational variables [were] comparable to those of females in the sample" (Heim and Estabrook, 1983, p. 37).

The good news is that since the 1980s, women have made significant strides in closing the managerial gap. Even so, they are still today underrepresented in higher administrative positions and still only earn approximately three-quarters of the wages earned by men (Golub, 2009).

There is also some suggestion that the prominence given to technology in libraries and the de-skilling of more traditional types of library work such as cataloguing and reference (particularly by passing these jobs down to less qualified library employees) are due to the increasing male influence in librarianship. Not surprisingly, men have attempted to distance themselves from the more traditional aspects of this female-dominated profession not only by changing the focus of the work, but also by changing the name of the profession itself. Once universally known as librarianship, professional schools now offer degrees in library science or information science (Golub, 2009).

These attempts to promote the image and status of the work by changing the focus as well as the name signal the ongoing problems with the profession's attempt to sell itself as a true profession. Yet the problems with image and status are real and ongoing as can be seen in the fact that the job ultimately lacks the three hallmarks of other professions: (a) the monopoly on the work in question, (b) the specific expertise and training which is impossible for others to learn without similar education, (c) the legal protection and sanctions regarding who can declare themselves to be that type of professional.

As mentioned above, the poor image of librarians may well be responsible for the unrelenting series of budget cuts and downsizing afflicting many libraries. As librarian Regina Powers (2011) writes,

> [o]n school visits, I ask what students think a librarian does. The response is always the same. "Librarians check out books. They read a lot. They tell people to be quiet." These misconceptions are held by adults too. When I told a friend that I was embarking on my graduate degree, he asked, "You need a master's degree in the Dewey Decimal System?"

So while shutting libraries may reflect poorly on organizations or elected officials, Powers notes, laying off library staff doesn't have the same effect. Those with the authority to do the budget cutting assume that "libraries can simply run themselves." With this attitude, Powers wonders, why not "replace us with phone trees, self-service checkout machines and volunteers?" Sadly, in both North America and the UK, this is exactly what is happening (O'Connor, 2012).

Poor management

As with any other type of organization, ineffective leadership is also a problem for libraries. The problem begins with schools of library and information science. Although librarians often end up working in supervisory and management positions, library schools do little in the way of preparing librarians to be effective supervisors and leaders. Not much is taught, for example, on how to inspire subordinates, how to negotiate effectively with stakeholders and superiors, how to handle tricky personnel issues, or how to be both a leader and a team player. Instead, the profession tends to be "passive, inert, and [to] drift along without giving proper attention to this extremely important topic" (Riggs, 2001, p. 16).

It seems hardly surprising, then, that studies in which library staff and library directors are asked to rank top leadership traits end up producing lists of qualities that differ significantly (Young et al., 2006). These results are more than merely a difference in opinion between younger and older workers; indeed, without a shared educational foundation in the practices and principles of management, librarians and their directors will probably not agree on which leadership qualities are most important. Lack of agreement, however, may be indicative of gaps in understanding; librarians will be less likely to comprehend the motives or appreciate the actions of their directors, and directors will be less likely to sympathize with the concerns of subordinates, adjudging them to be too removed from the strains of management to understand the key issues and concerns.

Insufficient training in managerial methods, however, does more than merely lead to gaps in understanding; it is also responsible for ineffective or dysfunctional management practices down the road. This ineptitude frequently manifests itself in a number of ways. Library leaders may be unwilling, for example, to consult with librarians, library staff and other key stakeholders (Staninger, 2012). They make poor decisions because they do not seek the advice of those with expertise in the organization. Where consultation happens or advice is sought, the act is often merely *pro forma* and input derived from such consultation does not alter the course of management decision-making.

In addition to lack of consultation, dysfunctional management practices also include micromanagement, inflexibility and the unwillingness to share control (Staninger, 2012). Complicating matters is the fact that certain types of libraries (e.g., public libraries) have a board or commission which oversees library management. These boards and commissions can make matters worse if they have a tendency to micromanage (Bennett, 2004) or, conversely, to ignore problems entirely ("Better Oversight," 2012). Micromanaging boards can stymie the efforts of competent managers to run the library effectively, while neglectful boards allow bad managers to wreak havoc for extended periods of time. All of this, combined with stressful economic times and with the downsizing of budgets and staff, can result in enormous problems for library management, and a stressful work environment for staff.

Budgets and staffing

In 2012, the American Library Association released an "infographic" (the equivalent of a poster presenting facts and statistics in visual form). The infographic reported that in that same year, 57 percent of US public libraries reported decreased or flat library budgets, and 40 percent of states reported decreased public library funding. The decrease in library funding inevitably meant a decrease in services and, almost as often, a cut in library staff either through lay-offs or by closing positions left vacant as a result of turnover or retirement. As one librarian surveyed by *Library Journal* in that same year commented,

> [t]he worst impact on the library in terms of budget cuts is in the area of staffing . . . We are constantly short-staffed and must do the work of two librarians at all times. This is our new normal. (Kelley, 2012)

Newspapers in recent years have been replete with stories about the "new normal": budget and staff cuts in school, public, academic, national and other libraries. A 2010 *Library Journal* survey, for example, found that 43 percent of respondents had experienced staff cuts. Libraries serving larger populations were hit hardest – 93 percent of libraries in areas with over 1 million people had lost staff. In addition, 46 percent of survey participants were now relying more than ever on volunteers and self-service devices, while 59 percent of libraries trained employees to do multiple jobs in order to deal with staff shortages (O'Connor, 2012).

The new normal also involves closing branches entirely and firing librarians. In 2009, for example, the directors of the Carnegie Library of Pittsburgh decided to shutter four of its 19 branches and merge two other branches. The Pittsburgh system was 114 years old and had never before in its history closed a branch ("Sad Chapter," 2009). Likewise, in 2012, the Huffington Post reported that 58 out of 124 District of Columbia schools would be re-opening without librarians in the next school year (up from 34 schools the year before). A spokesperson for the public schools noted that "[i]n these tough budget times, we have to make tough budget choices" (LaJoie, 2012). Cutting out the librarians was one of those tough choices, but the District of Columbia was not alone in its pain. Indeed, it was only one of many "hot spots" for library staff cuts in the country. Other states with similar problems, the article noted, included California, Arizona, Ohio and Michigan.

In California, for example, hundreds of school librarian positions and 328 full-time public library positions were cut in 2010 (Powers, 2011). Indiana, too, experienced massive cuts. Union representatives for the Gary Public Library announced that impending cuts would probably mean the end to their labour union. The president of the union's local noted that, in addition to complete job loss for some staff, others had been cut from full-time to part-time status leaving them ineligible for union membership. She herself was one of the casualties noting that "her salary [was to] drop to $10,000 from $24,000 and she [would] have no benefits after 32 years on the job" ("Gary library," 2012). The reason given for the budget crisis was insufficient revenue for public services as a result of a statewide cap on property tax.

Florida was also hit hard in 2012. In Jacksonville, the public library system was cut by more than 12 percent (Stepzinski, 2012) resulting in staff reductions. In the same year, Governor Rick Scott cut $1.5 milllion in funding from Florida's five regional library consortia which had been key in offering local libraries training, participation in a statewide interlibrary loans program, and access to ebooks (Kitzmiller, 2012). The Miami-Dade Public Library's operating budget for 2012 was cut by 30 percent; 153 part-time workers were chopped and full-time staff were reduced by 25 percent (Kelley, 2012).

But the pain experienced by libraries was not confined to the United States; indeed, library staff cuts also made headlines across the ocean in the UK. In 2012, library staff in North Somerset were informed of budget cuts which would mean the end to employment for 15 percent of the workers (Pickstock, 2012). Alarmed by the news, local residents in some villages offered to act as volunteer workers in hard-hit libraries rather than see them close.

That same year, *The Sentinel* newspaper reported that Stoke-on-Trent had closed certain libraries completely and cut the librarian complement by 43 percent (from 16 to 9). At the same time, full-time library positions in the Staffordshire County libraries had been cut by 17 percent (from 301 to 249), and in Cheshire East, five library positions had been eliminated (Campbell, 2012).

These library cuts in the UK were not unusual. The Chartered Institute of Library and Information Professionals (CILIP) released a report in 2012 based on a survey of 93 out of the 174 library authorities located in England, Wales and Northern Ireland. The Institute reported that 21 libraries had closed and 2,159 library workers or 10 per cent of all library staff had lost their jobs in the last year. Budgets had been cut by an average of 7.5 percent (about £39 million) during the same time period. Also, as with the library in North Somerset above, "13 per cent of authorities had set up community-managed libraries in order to keep local services going" (Hall, 2012). Local residents, it seems, would be required to pick up the pieces and volunteer to work for no salary if they wanted continued access to a local library.

In many parts of the globe academic libraries, too, have had their budgets slashed. While much of the cutting has been done without fanfare, in some places it has made the headlines. In 2011, newspapers reported that Melbourne University was cutting 30 library positions (Collins, 2011). A spokesman for the University noted that automation and digitization were responsible for the downsizing.

In the US, newspapers reported that University of California, Berkeley's library had lost 70 employees in the previous four years as a result of funding cuts and was now required to shed 20 more positions as a result of further state cuts (Brown, 2012). The library system was at such a loss over how to handle this news that it had turned to the local community for suggestions on how next to proceed.

In both the US and Canada, government libraries have also been hit by budget and staff cuts. In July 2011, Governor Rick Perry "signed a new state biennial budget (FY12/13) that reduced state funding for the Texas State Library and Archives Commission by 64 percent" (Kelley, 2012). In May 2012, more than 400 employees at Canada's national library (Library and Archives Canada) received notice that they could be among the 20 percent of employees expected to be cut over the next three years. Librarians at various Canadian government departments were also expected to be cut and their libraries closed.

Interestingly, when asked about the problem, Canadian Heritage Minister James Moore said there was

> "no need for concern" regarding cuts at Library and Archives. "I think half of the job losses that are happening, they were actually going to happen regardless of the federal budget. This has to do more with the modernization initiative that was launched some time ago." (Kirkup, 2012)

Clearly, the idea that cuts were the result of "modernization" rather than budget woes has been no great consolation to those staff members receiving pink slips.

As these various accounts indicate, libraries of all stripes have found themselves cutting staff over the last few years; and these were cuts made over and above a continual process of downsizing that had already been underway for decades.

Finally and not surprisingly, no sooner had libraries began to report problems than library associations also found themselves in trouble. In 2009, for example, the American Library Association reported that it was suspending the publication of its print "ALA Handbook of Organization" for 2010 and 2011 and requiring its staff to take Memorial Day and the Fourth of July as unpaid holidays in addition to three additional furlough days during the first half of the year. The organization cited lower revenues from publication, subscriptions and conferences as part of the reason for its financial woes (Oder et al., 2009).

Budgets and acquisitions

In addition to staff downsizing, library budget cuts often mean cuts to acquisitions. Long before the 2008 recession, Atkinson (1993) commented, "[c]ertainly one of the most visible and highly publicized economic challenges facing the academic library at this time is the decline in the purchasing power of the academic budget" (p. 33). At the time, Atkinson wrote that the rising costs of publications, particularly serials, along with the new but growing costs of electronic sources were already putting pressure on acquisitions budgets. Where, for example, the price of library purchases was increasing by 6 percent a year but the library acquisitions budget was stagnant, real purchasing power was actually decreasing by 6 percent a year. Libraries found themselves trying to reduce spending on certain items simply to be able to afford much of the rest of what they had been buying all along.

Since the 2008 recession, libraries have been hit with significant cuts to budgets which have not only affected staffing but also the ability to purchase materials for the collection. In the UK for example, severe budget cuts were said to be responsible for slashing the number of new books purchased in Cambridgeshire libraries by half (Havergal, 2012). In 2009/10, that number amounted to 102,146 library items purchased; the following year, it dropped to 60,868, and in 2011/12 to 48,057.

In the US, Louisiana eliminated all state funding-aid to public libraries from its budget. Many libraries had relied on this money to help them buy books and other library materials. As one newspaper article reported:

> Director of the Audubon Regional Library Mary Bennett Lindsay told the Library Journal that state aid made up 10 percent of the library's budget. "I'm just going to pray," she said. "We'll just have to cut back on books and hope we get through. If our server goes down or the switches go down, it's going to have to come from somewhere. It's not going to come from utilities. We're barely paying people above minimum wage, so it's not going to come out of salary. We may have to cut hours." (Driscoll, 2012)

The Commissioner of Administration in Louisiana, who also serves as the governor's chief budget aide, blamed "tight budget times" and said that health care and education would have to take priority. He noted that local libraries could still be supported with local rather than state money, not mentioning the budget problems that local governments were themselves facing.

Libraries in Texas, like those in Louisiana, also lost state money. In 2009, the state gave the Montgomery County Memorial Library System $112,000, but reduced it to $86,000 in 2010. In 2011, the library was given no money whatsoever. What was even more difficult, though, was that the library system sustained cuts in local funding as well. As Jerilynn A. Williams, director of the library system noted, in 2008, the county had given the library system approximately $828,000 from the county for the purchase of new books. That number was cut to $500,000 in 2009 and to $250,000 in 2010. More cuts were made in 2011. "We're whiplashed – from both sides," she noted. "Basically, we took a 70 percent cut in our funding over the past 3 years . . . We are feeling the impact. Imagine trying to eat or breathe with only a third of what you had two years ago" (Peyton, 2011).

Sadly, these stories are only a small sample of the cuts to acquisitions budgets that have hit libraries everywhere. Tough budget choices by local, municipal, county, state and even national governments, in addition to schools and other organizations, have meant big hits for libraries in terms of what they can buy.

Ironically, the shift from print sources to electronic was once (and often still is) thought of a panacea for cost-saving. We explore this particular source of stress for library staff next.

The shift from print to electronic sources

Few libraries anymore have collections which are entirely print; instead, most libraries have a mix of electronic and print sources, and some libraries even appear to be doing their best to dispense with print altogether. In 2005, for example, Stanford University decided to build a new engineering library and, at the same time, to cut down on the amount of print on its shelves. Just how much it planned to cut down was startling: the new library opened in 2010 with 85 percent fewer books than it had had in the old library (Chen, 2010).

Apparently less startling on the surface, a recent survey revealed that 95 percent of academic libraries have ebooks and expect to spend approximately 20 percent of their total acquisitions budget on ebooks alone in the next five years (Miller, 2011). Public libraries are not far behind; as of 2011, 82 percent offered access to ebooks and expected to spend approximately 8 percent of their acquisitions budget on these electronic materials in the next five years (Miller, 2011).

Yet these changes are unprecedented in the history of libraries and they also raise a number of questions. With electronic materials, multiple users can access the same ebook or other document simultaneously, not only from within the library's walls but from any other location (including home) via their computers. One important question, then, is whether patrons will continue to visit libraries if they can access materials without leaving home. A recent survey on students and libraries found that 87 percent of respondents came to the library specifically to borrow books or to browse books on site. These results seem to "fly in the face of current library orthodoxy that if we add group study and [areas for] relaxing between classes the proportion appreciating the qualities of the physical space increases significantly. The big question of course that needs asking is whether the provision of e-books will trigger the flight of students from the physical library space?" (Nicholas et al., 2008).

An obviously related question is: will libraries themselves become unnecessary middle men once publishers find some way to offer ebooks directly to patrons? Only recently, a number of publishers – Macmillan Publishing, Simon and Schuster, Penguin Group, Brilliance Audio, Hachette Book Group – refused to sell or license their ebooks to public libraries (Librarian in Black, 2012), confident that they could make more money offering their ebooks to readers directly. Other publishers have allowed libraries to purchase their ebooks but only on the proviso that the electronic book might only be loaned to one patron at a time (the same as print), or only loaned 26 times at most before a new (electronic) copy had to be purchased (Annoyed, 2012). Both conditions undermine the key advantages of electronic materials, i.e., that they can be used by multiple users at the same time, and that they never wear out and do not require repurchase.

In addition to the problems of long term survival and how to deal with publishers under the new model, libraries have had to worry about a number of other issues as the result of the growth of electronic information. These are discussed below.

a) Packaged content

The movement from print to electronic materials has given unprecedented power to publishers. In the print world, for example, publishers rarely, if ever, required libraries to buy a variety of other titles in hardcopy in order to obtain the specific title they desired. This is not true in the electronic world. There, publishers package and re-package materials requiring libraries to buy numerous items just to get the handful of titles they really want. As the Annoyed Librarian, columnist for Library Journal.com, has predicted,

> [w]e'll have a situation where libraries are useful only as cash cows for publishers, and content is controlled, organized, and made available only as the publishers wish. Forget about selection, because it won't be possible anymore. Libraries will take the packages of books on offer, or they won't. Publishers will realize that there's no point in pretending to sell individual books since they're just licensing content now. They'll be doing the selection for libraries, take it or leave it. There will be ebook packages based on obscure categories whose main purpose is to make money. There will be "academic" and "public" packages, but with enough missing from each that libraries will have to buy both to have even remotely comprehensive collections. There will also be current files and back files and every other possible way of dividing up the available books to make the most money from them. No matter what libraries try to do, they'll end up paying for a lot of junk they don't want so they can get the bit they do want. (Annoyed, 2010)

Libraries first encountered this problem when electronic journals appeared on the scene. Even though older hard-copy issues sat on the shelves, publishers often required libraries to buy the same titles again as part of an electronic package which included access to recent issues. As a result, libraries had electronic access to the same title through multiple packages for which, of course, they had already paid many times.

The same was also true for ebooks. In order to have access to current titles from publishers, libraries found themselves paying for electronic access to titles which they didn't want in the first place or which they had already purchased in hardcopy.

But having paid for access doesn't necessarily mean that access is predictable or reliable. In January 2012, for example, the University of Toronto (U of T) Press unilaterally decided to withdraw its content (about 600 titles) from an electronic collection called the e-Canadian Publishers Collection (one of three collections in the Canadian Electronic Library (CEL)) because of a dispute over how patrons were allowed to access these titles (University of Regina, 2012). Following the example of the U of T Press, three other university publishers – McGill-Queen's University Press, Wilfred Laurier University Press and University of British Columbia Press – also withdrew their content, and libraries everywhere who had subscribed to this service found themselves without electronic access to the titles issued by these publishers.

Across Canada, books in this package were being used by professors for their classes and subsequently libraries had to scramble to buy the material in hardcopy or find access in some other way. Library catalogues also had to be adjusted to reflect that these titles were no longer available. So, rather than creating less work for libraries, electronic materials often create more work along with a new type of chaos, specifically in situations such as this one.

b) Restrictions on use

In the past, libraries paid for a book which they then could lend over and over again to multiple patrons. The library's relationship with the book it had purchased was much the same as it would be with any other inanimate object – it could lend the book, re-sell the book, destroy the book or give it away. As a result of the "first sale doctrine" established by the US Supreme Court in 1908 in *Bobbs-Merrill Co. v. Straus*, 210 US 339 (and roughly equivalent doctrines in various other legal jurisdictions), individuals or organizations purchasing a book did not require permission from the copyright holder to sell or otherwise dispose of the book after purchase. Indeed, because of the robust market for used books, publishers had to remain somewhat reasonable in terms of the prices they charged (Spalding, 2009). Only where publishers had a true monopoly over certain titles, (e.g., certain titles published in academia) could their terms of sale become unreasonable by, for example, charging exorbitant prices for these materials.

All that has changed. Unlike print, the purchase of electronic information generally requires that libraries sign a contract which, among other things, contains restrictions on how the material may be used. For example, electronic books and journals generally come with restrictions on inter-library loan, transfer and re-sale. The restriction on inter-library loan is particularly worrisome in that libraries could find themselves in virtual silos unable to borrow materials from one another.

Yet, above and beyond the prohibitions on use in these contracts, the platforms on which the ebooks and journals exist often have restrictions on use built into the software. This means that users may be prohibited from printing, copying, pasting, downloading, emailing, or even moving the information to another computer. As the University Leadership Council (2011) report recently noted, "[i]ronically, it is now easier to share physical books than electronic copies. Until licensed or 'fair use' access to the mass-digitized corpus is resolved, colleges and universities will be unable to begin replacing physical collections with digital access to scanned material" (University Leadership Council, 2011, p. ix).

c) Multiplicity of platforms

The multiplicity of platforms is yet another problem. Ebooks are still read and searched on a variety of platforms with no one standard emerging yet from the pack. This makes it difficult for the library to troubleshoot technical problems. It also requires patrons to reorient themselves each time they encounter a book on a new platform.

The extent to which one is allowed, for example, to download, print, cut or paste depends on the particular platform being used. Some platforms allow for no printing or downloading; others allow for limited page printing but no downloading. Some permit cutting and pasting text; others do not. Trying to work out which is which is often confusing for patrons and maddening for library staff.

The layout of these platforms is also a problem. Often command bars fill much of the screen. It is not uncommon to read an ebook through an interface which has a command bar at the top of the screen, at the left hand side of the screen and at the bottom. The space left for the actual text shrinks dramatically and readers are often left feeling that they are reading a book through a keyhole.

d) Cost

Cost is yet another problem. The brave new world of electronic materials has also ushered in with it a new approach to pricing and access. Libraries have become used to the "cable TV" approach to payment in which they paid yearly for ongoing access, aware that if they stopped paying, access would disappear. They are also used to the "perpetual access" model.

Perpetual access involves a payment in which a library, in theory, would pay a price and presumably have access thereafter "forever." The library would either "host" the electronic material for which it had paid (i.e., it moved the contents to its own computers) or it paid an annual "hosting fee" to the publisher which would continue to have the information accessible from its computers.

Libraries were not ready, however, for the new model introduced by HarperCollins in 2010 which set out that "new titles licensed from library ebook vendors [would] be able to circulate only 26 times before the license expire[d]" (Hardo, 2011). Nor were they prepared for the host of other fees that would accompany the purchase of ebooks. As David Stern noted,

> [l]ibraries traditionally paid for books as one time purchase or as parts of standing order sets. There were no platform fees, aggregator fees, third-party support costs or annual maintenance fees. In many systems, there is no way to show and account for such continuation commitments. There are also few options to handle and reflect multiple seat charge options, revised charges based on use data, reserves use fees (as a short-term option or as a permanent right), or storage of use data and continuation decision histories. (Stern, 2010, p. 31)

Such is the new model emerging for libraries, and publishers in this model have an opportunity to increase their revenues like never before. At the same time, they are aware that they can't stay in business if they charge more than their customers can afford. Libraries which don't have the money simply won't buy or subscribe. So, as stressful as it seems, there will be much more experimenting with pricing before some standard finally emerges.

Nonetheless, even these irritants involving electronic materials, access and cost may seem trivial to library workers when compared with other problems facing them today. These range from quite unpleasant, low level nuisances to an increasing possibility of violence directed at information professionals on the job. It is to this particularly worrisome and stressful aspect of our work situation that we now turn.

Attacks, assaults and "low level nuisances"

Libraries are increasingly becoming unsafe places to work and staff, particularly those who work in libraries open to the public, have endured an unprecedented rise in physical and verbal attacks and assaults over the last few decades. In Derbyshire, England in 2010, for example,

> there were 93 assaults on staff in the council's cultural and community services department, most of which were on library staff. These included 13 physical attacks, one racial physical attack, three sexual physical, 73 verbal and one sexual verbal. This was more than treble the 2009 figure of 30. The figures for 2011 look set to be even higher with 42 assaults reported between January and March – 12 physical and 30 verbal. In 2010 the council banned 37 people from its libraries – up from 18 the previous year. (Mallett, 2011)

In other industrialized countries, stories of assaults on library workers abound. In 2009, in a small village library in the Ottawa, Canada region a 15-year-old library patron waited one evening until the library was empty, and then jumped on the lone employee's desk, beating her until she almost lost consciousness. She was then tied up, terrorized and repeatedly sexually assaulted. The patron then fled and was later apprehended by police. The library employee eventually recovered from her injuries; however, once the court proceedings were underway, her lawyer noted that she felt ashamed and embarrassed and didn't feel she could return to work given the sexual nature of her attack (Seymour, 2010).

In 2010, in the United States, the union representing employees at the Mount Vernon Public Library in New York took their employers to task citing a number of violent incidents in the system and the lack of response by administrators. These incidents included a violent assault on the library security guard who had an eye socket fractured (McCormack, 2010), and library employees who had books thrown at them or who were groped (Garcia, 2009). Also in the US, in 2010, a library worker in Dallas, Texas was sexually assaulted before the library opened, and in 2012, a children's librarian in Auburn, Maine was attacked by a 19-year-old patron who was described as having a "psychotic break." She was beaten and kicked repeatedly until he was restrained by another patron and library staff members (Skelton, 2012).

Only a handful of the examples of violence against library employees make the news. The vast majority, sadly, are never reported. But what they indicate is that working in libraries which are open to the public, i.e., public libraries or even libraries in post-secondary educational institutions, puts employees at greater risk for workplace violence. This is something understood by experts who focus on workplace health and safety matters.

The Canadian Centre for Occupational Health and Safety (2012), for example, lists several work-related factors that increase the risk for workplace violence. Among these are working with the public, handling money, providing a service, working with "unstable or volatile persons," working alone or with only a small number of people, working in the evening or early in the morning, working in an isolated part of the building (e.g., in the stacks), and so on. All of the above are conditions in which library staff work on a daily basis, and all of them put workers at greater risk of violent incidents.

It doesn't help, of course, that librarians and other library staff are given either no or inadequate training on how to handle workplace risks. Library schools generally do not offer training in workplace violence (McCormack, 2010). While violence and harassment clauses are fairly standard in library policy manuals, it is rarer to see these clauses followed up by extensive training.

Administrators are not always helpful either. For example, after the incident at the Auburn Public Library described above, the president of the Auburn Public Library Board of Trustees said that while the Board would review security procedures, the incident seemed like a random one and unlikely to repeat itself. "We like to offer the library as a safe and welcoming place, not security-ridden," he said (Skelton, 2012). Clearly, this kind of politic, evasive response is rather less than front-line library workers would have anticipated or desired following the assault.

Over and above incidents of out-and-out violence, however, library staff also have to endure a steady stream of "low-level nuisances" including "people engaging in noisy, disruptive behaviour and inappropriate language, then swearing at staff when asked to stop the behaviour or to leave the library" (Mallett, 2011). Library policies often contain information about how to handle things such as verbal abuse, personal hygiene complaints, inappropriate behavior, lurking, and not abiding by the rules generally. Trying to enforce these policies, though, put library staff in the uncomfortable position of having to play security guard – a job they were never trained to do.

They also have to cope with the reality of patrons preying on each other. In England in 2011, a 14-year-old girl with learning difficulties was sexually assaulted as she used a computer in a Bolton library. Assaults of this sort in the UK are becoming more common.

But patrons preying on other patrons are also a problem in the United States. In 2008, for instance, a teenager at the Bloomingdale Regional Library in Valrico, Florida was raped, strangled, and beaten so severely she lost her eyesight in addition to the ability to walk and talk (Nguyen, 2008). The rapist was another library patron.

More recently, a convicted sex offender was sentenced to life for sexually assaulting a 6-year-old boy in a Houston Library bathroom ("Convicted sex offender," 2012). Around the same time, a Butte, Montana man was convicted of holding a weapon to the throat of a Montana Tech student in the library restroom and then punching her in the head several times ("Butte man," 2012). In Orange County, California, another registered sex offender was charged with assaulting a boy in the library in March 2012 ("Registered sex offender," 2012). Meanwhile, a New York man was stabbed in the neck as he used a computer in a Brooklyn library ("Police investigate," 2012). Once again, these stories were only some of a number of reported assaults in the United States in which one library patron had been found preying on another in that year.

Conclusion

As the various issues discussed above indicate, librarianship is no longer a profession that is "a breeze" (Wilson, 2009). Poor status, massive budget cuts, staff downsizing, and branch closings have left many librarians working in an environment filled with constant anxiety and dread.

As if this weren't enough, the job itself is changing radically as libraries reduce their print content and move to collections that are largely electronic. Those who embraced librarianship because they loved books find themselves reeling at the wholesale dumping of print and the race to embrace electronic resources despite the obvious problems that include access, readability, pricing models, usability and so on.

As an added stressor, more and more information professionals are dealing with increasing levels of violence in the workplace. Those who are lucky enough to have avoided violent incidents have certainly been exposed to low level nuisances such as inappropriate behavior and destruction of library materials, often on a daily basis.

With all this in mind, it seems less surprising to hear that, of five specific jobs surveyed (firefighters, police officers, train operators, teachers and librarians), librarianship was found to be the most stressful (BBC News, 2006). Workplace stress, low status, lack of decision-making authority, low-level tasks, and the lack of systems of support all worked together to make librarianship, at least in this survey, more stressful than apprehending criminals or fighting fires.

References

American Library Association (2012) Infographic: Weather the storm. Retrieved from *http://www.ala.org/research/plftas/2011_2012/weatheringthestorm.*

Annoyed Librarian (2010) Libraries giving books away, or not. *Library Journal.* Retrieved from *http://blog.libraryjournal.com/annoyedlibrarian/2010/08/02/libraries-giving-books-away-or-not/.*

Annoyed Librarian (2012) Be grateful publishers don't like you. *Library Journal.* Retrieved from *http://blog.libraryjournal.com/annoyedlibrarian/2012/02/20/be-grateful-publishers-dont-like-you/.*

Atkinson, R. (1993) Crisis and opportunity: Reevaluating acquisitions budgeting in an age of transition. *Journal of Library Administration*, 19(2), 33–55.

BBC News (12 January 2006) Librarians 'suffer most stress.' Retrieved from *http://news.bbc.co.uk/2/hi/4605476.stm.*

Bennett, D.L. (5 February 2004) Report alarms library board, commission staffers tell of bias, poor management. *The Atlanta Journal – Constitution*, JH4.

Better oversight of Detroit library needed: Editorial (6 January 2012) *The Detroit News*, A14.

Bobbs-Merrill Co. v. Straus, 210 US 339 (1908).

Brodnick, M. (2 October 2012) Will public libraries become extinct? *Forbes.* Retrieved from *http://www.forbes.com/sites/quora/2012/10/02/will-public-libraries-become-extinct/.*

Brown, S. (4 May 2012) UC Berkeley's library ponders its future in a poorer, wired age. *San Francisco Business Times Online.* Retrieved from *http://www.bizjournals.com/sanfrancisco/news/2012/05/04/uc-berkeley-library-ponders-future.html?page=all.*

Butte man pleads guilty to assaulting Montana Tech student in library restroom (16 March 2012). Associated Press Newswires.

Campbell, A. (14 May 2012) Library jobs go in £700k budget cuts; Cash-saving measures will "impact communities". *The Sentinel,* 4.

Canadian Centre for Occupational Health and Safety (2012) Violence in the workplace. Retrieved from *http://www.ccohs.ca/oshanswers/psychosocial/violence.html.*

CareerCast (2010) The 10 least stressful jobs of 2010. Retrieved from *http://www.careercast.com/content/10-least-stressful-jobs-2010.*

Chen, A. (23 July 2010) Stanford goes electronic with newest library. *Palo Alto Online*. Retrieved from *http://www.paloaltoonline.com/news/show_story.php?id=17653.*

Collins, S. (3 August 2011) Students protest at library staff cuts. *The Age*, 5.

Convicted sex offender gets life in prison for assault on boy, 6, in Houston library bathroom (22 June 2012) Associated Press Newswires.

Driscoll, M. (2 July 2012) Louisana eliminates state funding for libraries. *Christian Science Monitor*. Retrieved from *http://www.csmonitor.com/Books/chapter-and-verse/2012/0702/Louisana-eliminates-state-funding-for-libraries.*

Fergus, E. (3 December 2010) Fury over move to downsize libraries. *Evening Times* (Glasgow, Scotland), Section: HS-News.

Garcia, E. (12 December 2009) Union: New safety measures at Mount Vernon library inadequate. *Library Worklife*. Retrieved from *http://ala-apa.org/newsletter/2010/01/28/union-new-safety-measures-at-mount-vernon-library-inadequate/.*

Gary library employees say a staff reorganization, including job cuts, will kill their union (28 February 2012) Associated Press Newswires.

Golub, E.M. (2009) Gender divide in librarianship: Past, present, and future. *Library Student Journal*. Retrieved from *http://www.librarystudentjournal.org/index.php/lsj/article/view/129/230.*

Hall, J. (16 March 2012) Scale of library cut-backs revealed; More than 2,000 library staff have lost their jobs in the last year as local authorities have cut services, while shorter opening hours have reduced public access to libraries by 3,000 hours a week, a report has found. *The Telegraph Online*. Retrieved from *http://www.telegraph.co.uk/news/9146818/Scale-of-library-cut-backs-revealed.html.*

Hardo, J. (25 February 2011) HarperCollins puts 26 loan cap on Ebook circulations. *Library Journal*. Retrieved from *http://www.libraryjournal.com/lj/home/889452-264/harpercollins_puts_26_loan_cap.html.csp.*

Havergal, C. (19 September 2012) Number of new books for libraries slashed by half. Retrieved from *http://www.cambridge-news.co.uk/Home/Number-of-new-books-for-libraries-slashed-by-half-19092012.htm.*

Heim, K., and Estabrook, L. (1983) *Career Profiles and Sex Discrimination in the Library Profession*. Chicago, Illinois: American Library Association.

Kelley, M. (26 April 2011) Geeks are the future: A program in Ann Arbor, MI, argues for a resource shift toward IT. *Library Journal*. Retrieved from *http://www.libraryjournal.com/lj/home/890328264/geeks_are_the_future_a.html.csp.*

Kelley, M. (16 January 2012) The new normal: Annual library budgets survey 2012. *Library Journal*. Retrieved from *http://lj.libraryjournal.com/2012/01/funding/the-new-normal-annual-library-budgets-survey-2012/.*

Kirkup, K. (3 May 2012) Federal librarians brace for cuts. *The Edmonton Sun*, 35.

Kitzmiller, F. (22 May 2012) Library system dealing with budget cuts. *The News Herald*, Panama City, Florida.

LaJoie, E. (14 August 2012) 58 D.C. schools will reopen without librarians. Retrieved from *http://www.huffingtonpost.com/2012/08/14/dc-schools-librarians_n_1753871.html?utm_hp_ref=libraries-in-crisis.*

Librarian in Black (9 February 2012) Notice to publishers: Curse your sudden but inevitable betrayal. Retrieved from *http://librarianinblack.net/librarianinblack/2012/02/ebooksign.html.*

Lloyd, D. (2010) The death of the library: Read it and weep. *Politics Daily.* Retrieved from *http://www.politicsdaily.com/2010/06/05/the-death-of-the-library-read-it-and-weep/.*

Mallett, C. (5 September 2011) Teenage yob attacks bring terror to staff in our libraries; Workers falling victim to violence and sex pests; Frontline library workers left in fear by rising tide of attacks. *Derby Evening Telegraph*, 1, 2.

McCormack, N. (2010) Staying safe: Revisiting security measures in libraries. *Feliciter*, 56(2), 74–5.

Miller, R. (12 October 2011) Dramatic growth: *LJ*'s second annual ebook survey. *The Digital Shift Library Journal.* Retrieved from *http://www.thedigitalshift.com/2011/10/ebooks/dramatic-growth-ljs-second-annual-ebook-survey/.*

Morrison, E. (22 August 2011) Are books dead, and can authors survive? *The Guardian.* Retrieved from *http://www.guardian.co.uk/books/2011/aug/22/are-books-dead-ewan-morrison.*

Nicholas, D., Rowlands, I., Clark, D., Huntington, P., Jamali, H.R., and Ollé, C. (2008) UK scholarly e-book usage: A landmark survey. *Aslib Proceedings: New Information Perspectives*, 60(4), 311–34.

Nguyen, D. (6 September 2008) Assaulted teen may regain her sight; So far, the student attacked at a library has regained only about 10 percent of her vision. *St. Petersburg Times*, 1A.

O'Connor, S. (1 June 2012) Public and academic library closures in the US, UK, and Eire, Part VII. *Examiner.* Retrieved from *http://www.examiner.com/article/public-and-academic-library-closures-the-u-s-u-k-and-eire-part-vii.*

Oder, N., Albanese, A., and Lau-Whelan, D. (2009) ALA belt-tightening includes staff cuts, dip into reserve. *Library Journal*, 134(7), 14–15.

Peyton, L. (19 July 2011) Budget cuts book buying at library. Retrieved from *http://www.ultimatewoodlands.com/stories/254164-community-budget-cuts-book-buying-at-library.*

Pickstock, H. (12 June 2012) Library staff set to lose jobs as council cuts keep coming; Ten full-time posts under threat. *Bristol Evening Post*, 15.

Police investigate bloody attack inside public library (16 May 2012) Retrieved from *http://abclocal.go.com/wabc/story?section=news/local/new_york&id=8663608.*

Powers, R. (26 October 2011) California must value librarians: Libraries can't run themselves. *Los Angeles Times.* Retrieved from *http://articles.latimes.com/2011/oct/26/opinion/la-oe-powers-librarians-20111026.*

Price, L. (10 August 2012) Dead again. *New York Times.* Retrieved from *http://www.nytimes.com/2012/08/12/books/review/the-death-of-the-book-through-the-ages.html?pagewanted=all.*

Prins, H. and de Gier, W. (1992) Image, status and reputation of librarianship and information work. *IFLA Journal*, 18(2), 108–18.

Registered sex offender charged with assaulting teens in library and airport (14 March 2012) Associated Press Newswires.

Riggs, D.E. (2001) The crisis and opportunities in library leadership. *Journal of Library Administration*, 32(3–4), 5–17.

Sad chapter: Downsizing comes to the Carnegie Library (9 October 2009) *Pittsburgh Post-Gazette (PA)*, Section: Editorial.

Seymour, A. (28 October 2010) Teen gets 3 years for library sex attack; 17-year-old gets maximum youth sentence for 'chilling,' 'horrendous' assault. *The Ottawa Citizen*, C1.

Skelton, K. (24 April 2012) Auburn librarian recovering after random attack. *Auburn Sun Journal* (Maine). Retrieved from *http://www.sunjournal.com/news/lewiston-auburn/2012/04/24/auburn-librarian-recovering-after-random-attack/1185411.*

Spalding, T. (2009) Ebook economics: Are libraries screwed?" *Library Thing*. Retrieved from *http://www.librarything.com/blogs/thingology/2009/10/ebook-economics-are-libraries-screwed/.*

Staninger, S.W. (2012) Identifying the presence of ineffective leadership in libraries. *Library Leadership and Management*, 26(1), 1. Retrieved from *http://journals.tdl.org/llm/index.php/llm/article/view/5782/5815.*

Stepzinski, T. (24 July 2012) How Jacksonville library's proposed cuts would affect services. *The Florida Times-Union.*

Stern, D. (2010) Ebooks from institutional to consortial considerations. *Online Magazine*, 34(3), 29–35. Retrieved from *http://www.infotoday.com/online/may10/Stern.shtml.*

University Leadership Council, Education Advisory Board (2011) Redefining the academic library: Managing the migration to digital information services. Retrieved from *http://www.educationadvisoryboard.com/pdf/23634-EAB-Redefining-the-Academic-Library.pdf.*

University of Regina (2012) Ebooks withdrawal from the Canadian Electronic Library alert. Retrieved from *http://www.uregina.ca/library/about_contact/library_information/announcements_content/2012/246.html.*

Wilson, D. (18–19 July 2009) Enjoy life in the slow lane. *The Sydney Morning Herald*. Retrieved from *http://content.mycareer.com.au/advice-research/career/enjoy-life-in-the-slow-lane.aspx.*

Young, A.P., Hernon, P., and Powell, R.R. (2006) Attributes of academic library leadership: An exploratory study of some Gen-Xers. *The Journal of Academic Librarianship*, 32(5), 489–502.

4

Are information professionals burned out?

Research and opinion

Abstract: This chapter provides an overview of empirical studies on burnout conducted specifically on groups of information professionals over the years. It also includes opinions offered by authors who have reviewed empirical research in this area.

Key words: burnout, information professionals, librarianship, empirical research, opinion.

Introduction

Burnout is a problem whose effects have been studied in numerous professions, occupations and other population groups. Indeed, researchers in 2009 put the number of burnout studies at somewhere between 5,500 (Casserley and Megginson, 2009) and 6,000 (Schaufeli et al., 2009). Today, that number is substantially higher.

As an indication of the scope of these occupational studies, in 2012 alone, burnout research was done on teachers (Pishghadam and Sahebjam, 2012), doctors (Siu et al., 2012), nurses (van der Doef et al., 2012), clergy (Barnard and Curry, 2012), occupational therapists (Gupta et al., 2012), music therapists (Kim, 2012), flight attendants (Chen and Kao, 2012), correctional staff (Lambert et al., 2012), professional caregivers in long-term care settings for the elderly (de Rooij et al., 2012), working undergraduate university business students (Galbraith and Merrill, 2012), dental and medical students (Prinz et al., 2012), eating disorder treatment providers (Warren et al., 2012), and employees at a large public hospital (Hämmig et al., 2012), to name but a few.

In this plethora of studies, information professionals as a group have not been entirely overlooked. Indeed, over the years several studies have been done on librarianship. But in addition to those, opinions on burnout have also been expressed by other information professionals who have not had the benefit of any kind of hard data. As Fisher (1990) notes,

> [e]mpirical investigations are severely outnumbered by ideological proclamations from the believers. The faithful are of one voice: library life is stressful and significant numbers of librarians are burning out. They do not often make direct comparisons with other professions, but when they do it is to show that librarianship is at least as stressful as other callings. They do not have to prove any of this, they *know* it is true. (p. 229; emphasis in original)

Consequently, over the last many decades there has been some disagreement over whether librarianship is a stressful occupation. Certain sources, indeed, have called it one of the least stressful occupations; at the other end of the spectrum, others have declared it to be more stressful even than policing or firefighting.

The 2000 edition of the *Jobs Rated Almanac*, for example, rated 250 jobs from most stressful to least. Librarianship was ranked the ninth lowest in terms of its occupational stress levels (Sheesley, 2001) – such occupations as bookbinders and barbers were ranked higher. As Marcia J. Nauratil (1989) has noted,

> [l]ibrarianship is popularly viewed as a pleasant and non-stressful sinecure in a quiet booklined setting, with relaxing routines gently punctuated by towheaded tots arriving for story hour and by chats with gray-haired matrons searching for fresh mysteries. (p. 1)

It is a view of the profession which has not necessarily changed all that much over time, and is still held even by those who work in it. As recently as 2009, the Annoyed Librarian comments,

> [a]s with the weather, librarians complain about work rather than doing anything about it. There could be many reasons for this inaction, but one of them is probably a lack of entrepreneurial spirit. Entrepreneurs have to go hustle up work. It's stressful because there

> is no dependable paycheck. If they want to be paid, they have to go make work happen, which often means working very long hours on short term projects where people demand results or you don't get paid. If you're so stressed, try going out and changing your life and making a living in the considerably more cutthroat private sector. Oh, you don't want to do that? I don't blame you. Try dealing with billable hours or having your paycheck depend on meeting sales quotas or having to go out and actually *find* projects on a weekly basis. Some of my acquaintances would look at the complaint and laugh. "Oh, you have to catalog books *and* answer reference questions, and your boss is dumb? You poor baby!" (Annoyed Librarian, 2009; emphasis in original)

Library personnel who complain about stress might indeed be stressed out, the Annoyed Librarian concludes, but if so, it is not their workplace which has made them that way.

Yet, in surprising contrast to such skepticism, a 2006 study found that library work was the most stressful of five different occupations – firefighters, police officers, train operators, teachers and librarians. Librarians taking part in the study saw their work as repetitive and unchallenging. They saw themselves as having little control in the workplace and one in three suffered from poor psychological health (BBC News, 2006). One of the reasons offered for being more stressed than firefighters or police officers was that library workers were less likely than individuals in these other occupations to have support systems in place designed to help them deal with stress.

Research and opinion

If, over the years, this group of surveys has delivered largely equivocal results, so has other more empirical research. For example, Smith and Nelson's (1983) study of academic reference librarians found low burnout rates, but Nauratil (1989) asserted that their findings were not reliable given a less-than-stellar methodology and the relatively obscure survey instrument used.

A similar comment was made by Fisher (1990) who questioned the findings made by Haack et al. (1984) who concluded from their study that 14 percent of librarians were severely burned out, and another 28 percent showed signs of psychological stress. Fisher and others have suggested that such questionable results were the product of an obscure survey instrument, administered to a non-random sample of people who knew what was being measured. Indeed, the group surveyed were attendees at a conference talk on burnout. Fisher noted that Christina Maslach, the author of the well-known and well-respected survey instrument, the Maslach Burnout Inventory (MBI), had warned against just such a practice. For her, revealing to survey participants that the survey about to be administered had been designed to measure burnout would invalidate the results, since participants had already become sensitized to the issue.

Around the same time, Smith and Nielsen, the same authors who had only recently surveyed academic librarians, turned their attention to special librarians employed by corporations. For this survey, the MBI was employed, revealing that 47 percent had a high burnout score in at least one of the categories of emotional exhaustion, depersonalization or decreased personal accomplishment. Librarians in this study reported that, for them, the specific sources of burnout were lack of positive feedback and having no authority or opportunity to participate in setting library practices or policies (Smith and Nielsen, 1984).

A few years later, a study involving 547 public library reference librarians (Birch et al., 1986) found role conflict and role ambiguity were, in particular, problems on the job. Role conflict takes place when workers find themselves with seemingly incompatible roles (e.g., cutting budgets and improving service). Role ambiguity, in contrast, has to do with workers who are unclear on what is expected of them on the job, often because their roles are ill defined. Birch et al. found that 35 percent of librarians were experiencing burnout linked largely to role conflict and role ambiguity. When burnout rates of librarians were compared with a study of teachers, they were found to be quite similar.

Law librarians seemed to fare noticeably better in a study published one year later by Nelson (1987). He surveyed a group of this particular subset of librarians and found average scores on the MBI in the moderate range. Fourteen percent had scores which indicated emotional exhaustion; score levels indicating depersonalization were found to be only slightly higher. But the scores indicating the extent to which librarians felt a sense of personal accomplishment were noteworthy: they indicated law librarians were not experiencing a great deal of job satisfaction.

Another paper, published within a few years of the above study of law librarians, measured occupational stress and burnout among library media specialists. These specialists experienced greater rates of burnout than seen in a number of previous studies on information professionals; workload and time pressures were found to be the major cause of stress in this specific group (Fimian et al., 1989).

In that same period, M.A. Nauratil's *The Alienated Librarian* (1989), while not an empirical work, reviewed much of the existing literature to date on burnout. Nauratil's theory was that libraries were over-bureaucratized resulting in stress, alienation, lack of job autonomy and burnout. The demand for greater productivity in an era of shrinking budgets, along with technological change, shifting views on how the library is marketed, more non-librarian managers and increasing reliance on consultants were all, in Nauratil's opinion, escalating sources of stress. To cope with it, she noted that worker autonomy and greater participation in policy and other decision making were vital in keeping burnout at bay. The ability to participate in this regard was also a hallmark of being a professional.

Following Nauratil's seminal work, Fisher (1990) surveyed the literature on stress and burnout and found only a handful of empirical studies, along with many opinion pieces. The empirical studies, according to Fisher, failed to make a strong case for any significant problem with burnout in libraries. "Many of the works we have analysed have turned out to consist of speculation and assertion rather than hard evidence." He continued,

> [t]hree articles have been based upon empirical research: one showed that academic librarians experience little or no stress; another that the majority of academic librarians surveyed suffer from a significant amount of stress/burnout; and the final piece discovered that the average public reference librarian experiences medium levels of burnout (a level shared by teachers). Hardly conclusive proof one way or the other. (p. 223)

Fisher also surveyed the anecdotal literature, but did not think highly of it. He concluded that ultimately, neither the empirical research nor the anecdotal literature added much of worth to our understanding of the problem and that more serious investigations were in order.

As with Nauratil's *The Alienated Librarian*, Janette S. Caputo's (1991) groundbreaking book, *Stress and Burnout in Library Service*, made no claim to be an empirical study in itself, but rather a survey of the library and social science literature regarding burnout. Caputo is clearly more impressed with the research to date than Fisher, noting:

> [w]ith one exception, surveys of librarian burnout to date suggest that this phenomenon is a very real problem for librarians, worthy of attention and concern. Most of the chronic work stressors experienced in the library field are similar to those experienced by other helping professionals. There are also some additional library-specific stressors that can have an impact on active librarians . . . it is easy to visualize the progress of a burning-out librarian through stages of enthusiasm, competency, stagnation, frustration and apathy. (p. 84)

Caputo went beyond merely summarizing the existing literature, however. Her book offered information on the causes of burnout, the signs and symptoms of burnout, and how to self-diagnose. In her later chapters, she went so far as to provide advice on how to cope with and prevent burnout.

A few years later, Becker (1993) examined a group of academic teaching librarians to discover whether they were exposed to factors which put them at risk for burnout and whether they were, in fact, experiencing any of the symptoms indicative of burnout. She determined there was indeed a clear risk for burnout among this group of librarians, which mirrored the difficulties faced by professionals in other helping professions.

Pursuing a similar line of research, Affleck (1996) used the MBI to investigate burnout levels in 142 bibliographic instruction librarians at comprehensive or liberal arts institutions in New England. Affleck found high levels of burnout in at least one of the three dimensions of burnout (emotional exhaustion, depersonalization, and diminished sense of personal accomplishment) for 52.8 percent of these bibliographic instruction librarians. She also discovered that 8.5 percent of them had scores which indicated burnout in all three areas.

Not long afterwards, Shtern (1999) used a survey based on the MBI to measure burnout rates among librarians in five Israeli university libraries –Bar-Ilan, Tel-Aviv, Haifa, Ben-Gurion and the Hebrew University in Jerusalem. The study found that the level of burnout fluctuated from low to medium, but overall indicated that it was relatively low among this population. The primary source of stress for Israeli academic librarians, Shtern discovered, was the profession's low status and the gap between the level of education required to do the job and the lack of remuneration consistent with the level of education. Librarians also were stressed by how little most people knew about the profession and its complexity.

Back in the US, Sheesley (2001) surveyed the literature on sources of stress and burnout in librarianship, particularly with regards to teaching. She noted that library work can be routine and boring, with little intellectual stimulation, and this included the teaching which is sometimes part of library work. Librarians who teach, Sheesley explained, are often asked to teach the same subject again and again, year after year, and to multiple sections of the same course. The repetitive nature of such teaching can be stressful, she asserted, as is dealing with students who are often less than enthused about taking instruction on library sources and research methods.

These librarians who teach, according to Sheesley's survey, often teach one or a handful of classes as part of another instructor's course (e.g., where a professor is assigning work that involves the use of library resources). They might also teach one-off courses for library patrons as part of the instruction offered by libraries for their patrons. The result of this type of teaching was that librarians missed the joy which came from getting to know students well and watching a class evolve and grow over the course of weeks or months. Teaching for others, or teaching one-off classes meant that librarians generally did not see the fruits of their own labor.

In addition, not only were librarians who teach not given the respect that is accorded to full-time instructors, they sometimes felt resented by some of their own colleagues or administrators who were of the opinion that teaching should not be part of the library's agenda. As a result, the teaching librarian was often made to feel that she was neither really part of the faculty nor really one of the staff.

Harwell (2008), another academic librarian, surveyed the literature and discussed his own personal experiences with burnout. Libraries, he noted, shared many of the same sources of stress as various other occupations. These sources included

> budget cuts, frequent technological changes, increased competition for fewer positions, heavy workloads, bureaucratic inertia, red tape, poor management and supervision, low pay, sex discrimination, obnoxious customers or patrons, lack of private work space or office, few opportunities to participate in goal setting or decision making, shifting priorities, lack of closure on ongoing projects, few opportunities for advancement, and working nights and weekends. (p. 382 citing Caputo, 1991)

Library jobs involving repetitive tasks or lack of stimulation could also be stressful. To this, Harwell added his own catalogue of personal stressors:

> [i]n my career, I have encountered numerous library deanships, difficult managers, temporary administrators, a rigorous tenure process, disappointing budgets, prolonged staff vacancies, redrawn mission statements, restructured organizations, innumerable changes in policies and practices, explicit position changes, and dozens of implicit changes in my day-to-day responsibilities. I have faced conflicting expectations from faculty, students, friends, and members of the larger community. I have learned a new focus to my work at least four times, starting with government documents, then patents and trademarks, law, and my current work, which is business and economics. (p. 385)

Such unending sources of stress, in combination, have the appearance of a prescription for burnout.

But library professionals are not the only sufferers. Paraprofessional cataloguers in academic and public libraries in Florida also experienced stress due to a number of different sources, according to a study by McClellan (2011). Paraprofessionals were expected to work too quickly and found it hard to do their work without making mistakes. The quantity of information they were expected to master along with keeping up with all the changes imposed a good deal of stress on them. A majority felt they needed more training in certain aspects of the job. Finally, they felt they had to balance conflicting needs and requests from a number of different departments which added to the stresses of the job.

Finally, in one of the latest studies in this line of research and commentary stretching back three decades, Ajala (2011) conducted group and individual interviews of librarians and other library staff members at the University of Ibadan Library System, Nigeria. Employees discussed the various issues which caused them stress on the job including lack of sufficient personnel, the requirement to publish while also working full time, and the inability to be involved in any decision making regarding the work. All decision making, employees noted, was done at the management level. These patterns are familiar and universal.

Conclusion

The literature surveyed does show evidence of burnout in libraries. But taken as a whole, these studies fail to provide us with enough definitive evidence to be able to make general statements about where specifically burnout is occurring, why and what can be done about it. As Fisher (1990) noted several decades ago, so much of what we know and believe continues to be based on opinion rather than on any great body of empirical evidence. His call for a credible, empirical approach still holds. As libraries face greater budget cuts, and profound technological and organizational changes, high quality research to determine whether employees in the information professions are at an equal or greater risk for burnout than those in other human services professions is more vital than ever before.

References

Affleck, M.A. (1996) Burnout among bibliographic instruction librarians. *Library & Information Science Research*, 18(2), 165–83.

Ajala, E.B. (2011) Work-related stress among librarians and information professionals in a Nigerian university. *Library Philosophy and Practice*. Retrieved from *http://digitalcommons.unl.edu/libphilprac/450/*.

Annoyed Librarian (27 May 2009) Just how stressful is library work? Retrieved from *http://blog.libraryjournal.com/annoyedlibrarian/2009/05/27/just-how-stressful-is-library-work/*.

Barnard, L.K., and Curry, J.F. (2012) The relationship of clergy burnout to self-compassion and other personality dimensions. *Pastoral Psychology*, 61(2), 149–63.

BBC News (12 January 2006) Librarians 'suffer most stress'. Retrieved from *http://news.bbc.co.uk/2/hi/4605476.stm*.

Becker, K.A. (1993) The characteristics of bibliographic instruction in relation to the causes and symptoms of burnout. *RQ*, 32(3), 346–55.

Birch, N., Marchant, M.P., and Smith, N. (1986) Perceived role conflict, role ambiguity, and reference librarian burnout in public libraries. *Library and Information Science Research*, 8(1), 53–65.

Caputo, J.S. (1991) *Stress and Burnout in Library Service*. Phoenix, Arizona: Oryx Press.

Casserley, T.B., and Megginson, D. (2009) *Learning from Burnout: Developing Sustainable Leaders and Avoiding Career Derailment*. Oxford, UK: Butterworth-Heinemann.

Chen, Ching-Fu, and Kao, Ya-Ling (2012) Investigating the antecedents and consequences of burnout and isolation among flight attendants. *Tourism Management*, 33(4), 868–74.

de Rooij, A.H.P.M., Luijkx, K.G., Declercq, A.G., Emmerink, P.M.J., and Schols, J.M.G.A. (2012) Professional caregivers' mental health problems and burnout in small-scale and traditional long term care settings for elderly people with dementia in the Netherlands and Belgium. *JAMDA*, 13(5), 486.e7–486.e11.

Fimian, M.J., Benedict, S.A., and Johnson, S. (1989) The measure of occupational stress and burnout among library media specialists. *Library and Information Science Research*, 11(1), 3–19.

Fisher, D.P. (1990) Are librarians burning out? *Journal of Librarianship and Information Science*, 22(4), 216–35.

Galbraith, C.S., and Merrill, G.B. (2012) Academic and work-related burnout: A longitudinal study of working undergraduate university business students. *Journal of College Student Development*, 53(3), 453–63.

Gupta, S., Paterson, M.L., Lysaght, R.M., and von Zweck, C.M. (2012) Experiences of burnout and coping strategies utilized by occupational therapists. *Canadian Journal of Occupational Therapy*, 79(2), 86–95.

Haack, M., Jones, J.W., and Roose, T. (1984) Occupational burnout among librarians. *Drexel Library Quarterly*, 20(2), 46–72.

Hämmig, O., Brauchli, R., and Bauer, G. (2012) Effort–reward and work–life imbalance, general stress and burnout among employees of a large public hospital in Switzerland. *Swiss Medical Weekly*, 142. Retrieved from *http://www.ncbi.nlm.nih.gov/pubmed/22653680*.

Harwell, K. (2008) Burnout strategies for librarians. *Journal of Business & Finance Librarianship*, 13(3), 379–90.

Kim, Y. (2012) Music therapists' job satisfaction, collective self-esteem, and burnout. *The Arts in Psychotherapy*, 39(1), 66–71.

Lambert, E.G., Hogan, N.L., Barton-Bellessa, S.M., and Jiang, S. (2012) Examining the relationship between supervisor and management trust and job burnout among correctional staff. *Criminal Justice and Behavior*, 39(7), 938–57.

McClellan, E. (2011) Stress and cataloging paraprofessionals in academic and public libraries in Florida. *The Southeastern Librarian*, 59(1). Retrieved from *http://digitalcommons.kennesaw.edu/seln/vol59/iss1/5*.

Nauratil, M.J. (1989) *The Alienated Librarian*. New York, New York: Greenwood Press.

Nelson, V.C. (1987) Burnout: A reality for law librarians? *Law Library Journal*, 79(2), 267–275.

Pishghadam, R., and Sahebjam, S. (2012) Personality and emotional intelligence in teacher burnout. *The Spanish Journal of Psychology*, 15(1), 227–36.

Prinz, P., Hertrich, K., Hirschfelder, U., and de Zwaan, M. (2012) Burnout, depression and depersonalisation – psychological factors and coping strategies in dental and medical students. *GMS Zeitschrift für Medizinische Ausbildung*, 29(1), Doc10. Retrieved from *http://www.ncbi.nlm.nih.gov/pubmed/22403595*.

Schaufeli, W.B., Leiter, M.P., and Maslach, C. (2009) Burnout: 35 years of research and practice. *Career Development International*, 14(3), 204–20.

Sheesley, D.F. (2001) Burnout and the academic teaching librarian: An examination of the problem and suggested solutions. *The Journal of Academic Librarianship*, 27(6), 447–51.

Shtern, O. (1999) *Burnout among librarians in Israel's academic libraries: Scope, rate and reason* (MA Thesis). Bar-Ilan University, Ramat-Gan, Israel.

Smith, N.M., and Nelson, V.C. (1983) Burnout: A survey of academic reference librarians. *College and Research Libraries*, 44(3), 245–50.

Smith, N., and Nielson, L.F. (1984) Burnout: A survey of corporate librarians. *Special Libraries*, 75(3), 221–7.

Siu, C.F.Y., Yuen, S.K., and Cheung, A. (2012) Burnout among public doctors in Hong Kong: Cross-sectional survey. *Hong Kong Medical Journal*, 18(3), 186–92.

van der Doef, M., Mbazzi, F.B., and Verhoeven, C. (2012) Job conditions, job satisfaction, somatic complaints and burnout among East African nurses. *Journal of Clinical Nursing*, 21(11–12), 1763–75.

Warren, C.S., Schafer, K.J., Crowley, M.E., and Olivardia, R. (2012) A qualitative analysis of job burnout in eating disorder treatment providers. *Eating Disorders*, 20(3), 175–95.

5

How burnout is measured in the workplace

Abstract: This chapter explains how burnout is measured in the workplace. It provides an overview of several of the key methods of measuring burnout including the Maslach Burnout Inventory (MBI), the Burnout Measure (BM), the Copenhagen Burnout Inventory (CBI), the Oldenburg Burnout Inventory (OLBI), and the Shirom–Melamed Burnout Questionnaire (S-MBQ).

Key words: burnout measurement, Maslach Burnout Inventory (MBI), Burnout Measure (BM), Copenhagen Burnout Inventory (CBI), Oldenburg Burnout Inventory (OLBI), Shirom–Melamed Burnout Questionnaire (S-MBQ).

Introduction

Over the years, researchers have sought a way to measure levels of burnout either in individuals or in occupational groups. Putting together a reliable instrument has been a problem given the debate over the very definition of burnout. For researchers, it is a "chicken and egg" problem: how are they to measure the levels of a syndrome when they cannot agree precisely on what they're supposed to be measuring? On the other hand, let's imagine a researcher chooses a specific definition and puts together a means of measuring the syndrome accordingly. If the definition is not deemed acceptable to other researchers in the same field, what is the value of the results? Are they totally meaningless? Questions such as these have plagued the study of burnout since the earliest days of such research, and debates regarding the definition of the syndrome have been especially problematic for empirical research (Schaufeli et al., 1993).

Nevertheless, a variety of methods for measuring burnout have been developed, with emphases that accommodate differences in the researchers' understanding of what burnout is.

In the early 1980s, Professors Christina Maslach and Susan Jackson put together a means of measuring burnout – a "burnout inventory" – which has since become the most popular tool. Their inventory measures the extent to which workers in several types of human services occupations report their experiences of symptoms in three dimensions of burnout: emotional exhaustion, depersonalization (i.e., "negative, cynical attitudes and feelings about one's client" (Maslach and Pines, 1979)) and reduced personal accomplishment (i.e., "a tendency to evaluate one's own performance negatively" (Maslach and Pines, 1979)).

These three dimensions are not the product of hypothesis; they were derived from studies done by Maslach and Jackson on various human services occupations. Not unimportantly, in the years since the publication of the Maslach Burnout Inventory (MBI), research involving tens of thousands of individuals in the human services professions has clearly shown the persistence of the three dimensions it measures (Leiter and Schaufeli, 1996).

Some researchers not wholly content with Maslach and Jackson's original inventory have themselves published burnout measurement instruments. Like the Maslach and Jackson burnout inventory, these newer burnout-measurement instruments continue to "assess feelings and emotions that are generated in work-related settings" (Schaufeli et al., 1993). Also, like the MBI in one of its later iterations, other researchers have designed their survey instruments for use in non-human services occupational groups. What differs, however, is that these other burnout measurement instruments often use a slightly different definition of burnout than Maslach and Jackson.

What follows is an overview of some of the measurement instruments most frequently used to measure rates of burnout in the workplace and in other population groups. They are alike in that all of them contain sample statements called "items" (e.g., "I feel very energetic" or "Working with people directly puts too much stress on me"). Survey participants are expected to respond to these items by indicating the extent to which or the frequency with which they have experienced the sentiments mentioned in the inventory.

The dissimilarities among the instruments are related to what they attempt to measure. It is here that one is able to observe the variations in opinion in terms of what burnout is, and how one should define it.

Burnout survey instruments

Maslach Burnout Inventory

The MBI (Maslach and Jackson, 1981, 1986; Maslach et al., 1996) is the survey instrument most commonly used to measure burnout in various populations (Densten, 2001). The MBI as originally conceived contained 25 items spread over four subscales (emotional exhaustion, depersonalization, and feelings of reduced personal accomplishment and involvement).

A second version of the MBI, published in 1986, contained 22 items with three subscales intended to measure the following burnout components: emotional exhaustion, depersonalization, and feelings of reduced personal accomplishment. "Involvement" was omitted. The inventory was created for individuals who worked in human services fields. Higher scores on the first two components and a lower score on the third would indicate to the researcher moderate to severe degrees of burnout in the employee who had responded to the survey. A third edition of the MBI (also called the Maslach Burnout Inventory-Human Services Survey (MBI-HSS)) was published in 1996 and meant specifically for human services professions (Maslach et al., 1996).

As noted above, the MBI is intended to measure burnout specifically among individuals who work in human services professions with direct recipients of these services, i.e., clients, patrons, students, patients, etc. The twenty-two items (i.e., statements including "I feel emotionally drained from my work," "I feel frustrated by my job") are answered on a frequency scale ranging from "never" (an answer which is given 0 points) to "every day" (an answer given 6 points). The frequency scale is meant to sort out individuals who have "bad days" interspersed with "good days" from those who are experiencing certain feelings or thoughts much or most of the time (Maslach, 2011).

Variant versions of the MBI exist including a version for those who work in education (the MBI-Educator Survey (MBI-ES): Maslach and Jackson, 1986a). This survey's items specifically refer to students rather than the recipients of one's service, care or instruction (e.g., "I don't really care what happens to some students"). In the same manner as the MBI for human services professionals, the instrument consists of 22 items which are answered on a frequency scale ranging from 0 ("never") to 6 ("every day"). Once again, the labels for the three dimensions measured are emotional exhaustion, depersonalization and reduced personal accomplishment.

Another version, the MBI-General Survey (MBI-GS; Schaufeli et al., 1996), is intended to measure burnout among groups of individuals who do not have a significant human services component as part of their job. Questions focus specifically on the job rather than the impact of dealing with other people as an integral part of the job. Hence the items to be responded to are, for example, "I feel emotionally drained from my work;" "I have become less enthusiastic about my work;" "I can effectively solve the problems that arise in my work" (Schaufeli et al., 1996). Given that the occupations are not people-oriented, the labels for the three dimensions measured are exhaustion, cynicism and reduced personal efficiency.

Like the MBI for human service professionals, the MBI-GS measures exhaustion; however, unlike the MBI, the five items on this subscale do not contain questions which explore the extent to which other people are the source of the exhaustion. The second subscale, cynicism, contains five items which measure negative attitudes and whether an individual has distanced him or herself from work.

The third subscale, reduced personal efficiency, contains six items which measure the negative assessment of one's present and past accomplishments, as well as feelings regarding any decline in competence. Like the MBI for human services professionals, the items are answered on a 6-point, frequency scale ranging from 0 ("never") to 6 ("every day"). Unlike the MBI and the MBI-ES which have 22 items, the MBI-GS consists of 16 items.

The MBI and its various versions have been translated into several languages, and have been used to measure burnout in a variety of occupations in countries around the world. Researchers generally support the MBI's three subscales (Lee and Ashforth, 1990; Taris et al., 1999; Schaufeli et al., 2001; Schutte et al., 2000); however, some have continued to have their doubts. As a result, certain researchers in the burnout community have sought to develop alternative inventories. These alternatives to the MBI are discussed below.

Burnout Measure

The Burnout Measure (BM) is a 21-item questionnaire developed by Pines and Aronson to measure the "state of physical, emotional and mental exhaustion caused by long-term involvement in situations that are emotionally demanding" (1988, p. 9). It is the instrument used most frequently as an alternative to the MBI. The 21 items are rated on a

7-point frequency scale (with 1 = "never" and 7 = "always") in which individuals are asked about their feelings and symptoms. The authors of the measure use an algorithm to derive a score. Individuals who score a 4 or higher are deemed to be burned out. Unlike the MBI, the instrument may be used for all occupations and is also suitable for non-occupational groups, that is, groups sharing some characteristic other than doing the same kind of work.

The authors have tested the measure on thousands of subjects with varying backgrounds including those in human services occupations, as well as business, management, science, administration, technical jobs, clerical work, and teaching. The BM has also been used to measure burnout in students, and is considered a reliable and valid instrument (Schaufeli et al., 1993). Nonetheless, unlike the MBI discussed above which measures three burnout components (emotional exhaustion, depersonalization, and feelings of reduced personal accomplishment), the BM measures exhaustion only.

Copenhagen Burnout Inventory

The Copenhagen Burnout Inventory (CBI) is intended to measure fatigue and exhaustion as indicators of burnout. Unlike the MBI, it does not measure depersonalization and reduced personal accomplishment.

The CBI consists of three scales. The first measures personal burnout, and therefore can be used for everyone, even individuals who are not employed. The questions here have been influenced in part by the BM, but are worded differently and have different response options (Kristensen et al., 2005, p. 200). "How often are you physically exhausted" and "How often do you feel weak and susceptible to illness?" (Kristensen et al., 2005, p. 200) are just two of the six items in this section.

The second scale is influenced by the questions having to do with emotional exhaustion on the MBI/MBI-GS instruments. This scale measures work-related burnout and, therefore, is applicable only to those individuals who do paid work. The seven items on this scale include "Do you feel worn out at the end of the working day?" and "Is your work emotionally exhausting?" (Kristensen et al., 2005, p. 200).

The final scale measures client-related burnout and is meant specifically for those individuals who work with human services recipients such as clients, patients, students, patrons, or inmates. It does not measure work with customers or colleagues. The six items on this scale include "Does it drain your energy to work with clients?" and "Do you sometimes wonder how long you will be able to continue working with clients?" (Kristensen et al., 2005, p. 200).

The CBI is appropriate for jobs which have varying degrees of client contact. Researchers have also discovered that, in the various groups studied, the "three scales predicted future sickness, absence, sleep problems, use of pain-killers, and intention to quit" (Kristensen et al., 2005, p. 192). Clearly, this information is particularly useful to organizations experiencing higher rates of burnout, absenteeism and employee turnover.

Oldenburg Burnout Inventory

One complaint about the original MBI was that it was designed specifically to measure burnout in individuals working in human services fields, while another was that all the items in a particular subscale were worded in a similar way. So, for example, the exhaustion and depersonalization subscales were worded negatively ("I feel emotionally drained from my work," and "Working with people all day is a real strain for me") while the items in the personal accomplishment scale were worded positively ("I have accomplished many worthwhile things in this job," and "I feel I am positively influencing other people's lives through my work").

The Oldenburg Burnout Inventory (OLBI; Demerouti, 1999; Demerouti et al., 2003) was intended to rectify these two complaints. The OLBI measures the levels of exhaustion and disengagement from work in various types of workers including those who do primarily physical work and those who are involved in information processing. The exhaustion subscale contains 7 items intended to measure physical and mental overtaxing and the need for rest. Four of the seven items are worded negatively and the remainder positively; for example: "After my work, I usually feel worn out and weary," and "After my work, I usually feel totally fit for my leisure activities" (Demerouti, 2001, p. 503).

The eight items on the disengagement subscale, in contrast, are intended to measure whether and to what extent individuals have distanced themselves from their work and whether they have developed negative or cynical attitudes towards the work they do. Items on the subscale include "I usually talk about my work in a derogatory way" and "I get more and more engaged in my work" (Demerouti, 2001, p. 503). The answering categories for both subscales are the same: 1 = "totally disagree"; 4 = "totally agree."

The OLBI has been found to work in both the human services context and in other occupational settings in which employees do not work with human services recipients (Demerouti et al., 2002). In theory, this makes it somewhat more flexible (although, arguably, less specific) than those surveys geared towards a particular subset of the working population.

Shirom–Melamed Burnout Questionnaire

For Shirom, burnout is "the chronic depletion of an individual's energetic resources" as a result of exposure to occupational stress (Shirom, 1989, p. 33). The Shirom–Melamed Burnout Questionnaire (S-MBQ) consists of 22 items which measure levels of burnout by focusing on five dimensions: emotional and physical exhaustion, tension, listlessness and cognitive weariness (Kushnir and Melamed, 1992; Melamed et al., 1999). Unlike the MBI, the S-MBQ attempts to separate out coping mechanisms such as distancing oneself from clients and work, and the consequences of burnout (a decrease in performance whether real or perceived).

A recent study using the measure (Lundgren-Nilsson et al., 2012) found problems with the S-MBQ in its original 22-item form. A revised 18-item survey, in which items measuring tension were removed, worked well. Researchers concluded that the total score from the questionnaire was a sufficient statistic for screening for burnout.

Conclusion

Despite its occasional detractors, the MBI remains the survey instrument used most often to determine the presence and the extent of burnout in specific populations. Nonetheless, the various other instruments discussed above, either in their original or revised forms, have also garnered supporters. Appropriately, those who study burnout will continue to tinker with such instruments in order to produce other and better ways of studying the syndrome with the goal of more precisely identifying both individual workers and larger worker population groups who are at risk.

References

Demerouti, E. (1999) *Burnout: Eine Folge konkreter Arbeitsbedingungen bei Dienstleistungs- und Produktionstätigkeiten.* (Burnout: A Consequence of

Specific Working Conditions among Human Service and Production Tasks). Frankfurt/Main, Germany: Lang.

Demerouti, E., Bakker, A.B., Nachreiner, F., and Schaufeli, W.B. (2001) The job demands–resources model of burnout. *Journal of Applied Psychology*, 86(3), 499–512.

Demerouti, E., Bakker, A.B., Nachreiner, F., and Ebbinghaus, M. (2002) From mental strain to burnout. *European Journal of Work and Organizational Psychology*, 11(4), 423–41.

Demerouti, E., Bakker, A.B., Vardakou, I., and Kantas, A. (2003) The convergent validity of two burnout instruments: A multitrait–multimethod analysis. *European Journal of Psychological Assessment*, 19(1), 12–23.

Densten, I.L. (2001) Re-thinking burnout. *Journal of Organizational Behavior*, 22(8), 833–47.

Kristensen, T.S., Borritz, M., Villadsen, E., and Christensen, K.B. (2005) The Copenhagen Burnout Inventory: A new tool for the assessment of burnout. *Work & Stress*, 19(3), 192–207.

Kushnir, T., and Melamed, S. (1992) The Gulf War and its impact on burnout and well-being of working civilians. *Psychological Medicine*, 22(4), 987–95.

Lee, R.T., and Ashforth, B.E. (1996) A meta-analytic examination of the correlates of the three dimensions of job burnout. *Journal of Applied Psychology*, 81(2), 123–33.

Leiter, M.P., and Schaufeli, W.B. (1996) Consistency of the burnout construct across occupations. *Anxiety, Stress and Coping*, 9(3), 229–43.

Lundgren-Nilsson, A., Jonsdottir, I.H., Pallant, J., and Ahlborg Jr., G. (2012) Internal construct validity of the Shirom–Melamed Burnout Questionnaire (SMBQ). *BMC Public Health*, 12(1), Retrieved from *http://www.biomedcentral.com/1471-2458/12/1*.

Maslach, C. (2011) Engagement research: Some thoughts from a burnout perspective. *European Journal of Work and Organizational Psychology*, 20(1), 47–52

Maslach, C., and Jackson, S.E. (1981) *Maslach Burnout Inventory. Research Edition*. Palo Alto, California: Consulting Psychologists Press.

Maslach, C., and Jackson, S.E. (1986) *Maslach Burnout Inventory* (2nd ed.). Palo Alto, California: Consulting Psychologists Press.

Maslach, C., and Jackson, S.E. (1986a) *Maslach Burnout Inventory – Educators Survey*. Mountain View, California: CPP, Inc.

Maslach, C., Jackson, S.E., and Leiter, M.P. (1996) *Maslach Burnout Inventory Manual* (3rd ed.). Palo Alto, California: Consulting Psychologists Press.

Maslach, C. and Pines, A. (1979) Burn-out: The loss of human caring. In A. Pines and C. Maslach (Eds.), *Experiencing Social Psychology: Readings and Projects*. New York, New York: Knopf.

Melamed, S., Ugarten, U., Shirom, A., Kahana, L., Lerman, Y., and Froom, P. (1999) Chronic burnout, somatic arousal and elevated cortisol levels. *Journal of Psychosomatic Research*, 46(6), 591–8.

Pines, A., and Aronson, E. (1988) *Career Burnout: Causes and Cures*. New York, New York: Free Press.

Schaufeli, W.B., Bakker, A.B., Hoogduin, K., Schaap, C., and Kladler, A. (2001) On the clinical validity of the Maslach Burnout Inventory and the Burnout Measure. *Psychology and Health*, 16(5), 565–82.

Schaufeli, W.B., Enzmann, D., and Girault, N. (1993) Measurement of burnout: A review. In W.B. Schaufeli, C. Maslach, and T. Marek (Eds.), *Professional Burnout: Recent Developments in Theory and Research*, pp. 199–215. Washington, DC: Taylor and Francis.

Schaufeli, W.B., Leiter, M.P., Maslach, C., and Jackson, S.E. (1996) The Maslach Burnout Inventory – General Survey. In C. Maslach, S.E. Jackson, and M.P. Leiter (Eds.), *Maslach Burnout Inventory Manual* (3rd ed.). Palo Alto, California: Consulting Psychologists Press.

Schutte, N., Toppinen, S., Kalimo, R., and Schaufeli, W. (2000) The factorial validity of the Maslach Burnout Inventory-General Survey (MBI-GS) across occupational groups and nations. *Journal of Occupational and Organizational Psychology*, 73(1), 53–67.

Shirom, A. (1989) Burnout in work organizations. In C.L. Cooper and I. Robertson (Eds.), *International Review of Industrial and Organizational Psychology*, pp. 25–48. Chichester, UK: Wiley.

Taris, T.W., Schreurs, P.J.G., and Schaufeli, W.B. (1999) Construct validity of the Maslach Burnout Inventory-General Survey. A two-sample examination of its factor structure and correlates. *Work & Stress*, 13(3), 223–37.

6

Burnout: the legal perspective

Abstract: This chapter discusses the law as it relates to burnout and stress in three jurisdictions: Canada, the United States, and the United Kingdom. Emphasis is given to issues surrounding workers' compensation, with less focus on occupational health and safety. In the United Kingdom context, there is also a discussion related to directives as mandated by the European Union.

Key words: burnout, stress, post-traumatic stress, mental injury, Canada, United States, United Kingdom, jurisprudence and legislation, workers' compensation, occupational health and safety.

Introduction

Inevitably, when a job makes an employee so ill as to suffer burnout, something has to give. A person may take sick leave, leave the job, or be fired. None of these options are particularly desirable, yet when burnout progresses to a point where an employee can no longer function, few other options exist. Even when an employer is willing and able to affect changes in the workplace, it may not be enough. Change in the workplace is only one part of the solution – while adjusting duties, responsibilities, and the reporting structure certainly helps, cultural change within the organization is also necessary. Without cultural change, a worker suffering from burnout is less likely to fully recover.

Nonetheless, an organization might only be willing to change once a burnout sufferer has taken legal action. In this case, what does the law actually say about burnout? What legal rights does a sufferer of work-related burnout have? What rights do employers have? Has there been any guidance from courts or governments on how workplace burnout must be handled?

In the three jurisdictions which we deal with here, there is little established law on the issue, making the problem that much harder to deal with effectively. There does, however, seem to be change on the horizon in jurisdictions such as the United Kingdom. Even countries such as Canada are making small concessions in the jurisprudence to employees suffering from burnout. Regardless of such positive developments, it seems clear that it will be some time before burnout as a specific mental injury sustained at work will be widely accepted as compensable in both the employment and legal spheres.

Before we delve into the law, we will first provide some context, describing both the essential differences between burnout and other stress-related disorders and the power imbalance in the work relationship.

Burnout, stress, or post-traumatic stress?

In Chapter 1, we saw that burnout and stress are related yet separate maladies. As Casserley and Megginson (2009) proffer,

> [s]tress is viewed as a temporary adaption to environmental stressors that brings with it psychological and physical symptoms while burnout indicates a failure to adapt resulting in "chronic malfunctioning". . . . [Burnout] also involves symptoms such as the manifestation of negative attitudes and behaviours at work which are not traditionally considered a stress variable. (p. 17)

While burnout and stress are distinct, North American courts and tribunals often understand them as being one. The same goes for post-traumatic stress disorder (PTSD), although that, too, is a distinct and separate ailment. PTSD "occurs after a traumatic event that involved actual or threatened death or serious injury, or a threat to the physical integrity of self or others" (Bisson, 2007, p. 789). Others describe it as "a psychiatric disorder caused by exposure to a traumatic event or extreme stressor that is responded to with fear, helplessness, or horror" (Mealer et al., 2009, p. 1118). Burnout and PTSD are quite dissimilar in their makeup, yet some in the legal field have treated them as though they are the same, or have mislabelled instances of burnout as PTSD. In the courts, burnout as a separate malady is generally rejected, at least in the North American context.

Power imbalance between employees and employers

It is well-established in law that an inequality of power exists between employers and employees (England, 2008, p. 3; Svirsky, 1998, p. 568; *Wallace v. United Grain Growers Ltd.*, 1997; *Slaight Communications Inc. v. Davidson*, 1989). In non-union environments, employers have an abundance of power, leaving workers at a disadvantage. This begins early in the relationship, starting with the job advertisement, which employers often describe as a competition. The disadvantage becomes abundantly clear with the initial negotiation of the employment contract (England, 2008, p. 44). Employers, with resources, power, and knowledge, may effectively wield the employment contract as "a vehicle to effectuate their real-world dominance over the employee" (England, 2008, p. 3). Employees rarely have the strength, confidence, or skills to negotiate effectively, especially during times when jobs are scarce. A person in need of employment often will not risk upsetting a potential employer by negotiating very hard (or at all).

Nonetheless, at certain times or in certain sectors, employees may actually boast more power due to an overall lack of workers in the field (England, 2008, pp. 105–6). For example, employers have recently complained that there are not enough skilled workers available to fill jobs in Canada (Blackwell, 2012). This desperate search for skilled workers has resulted in a call for increased immigration and spending on training and education (Blackwell, 2012). Such examples usually signal only a temporary shift in the power balance, however, and do not necessarily migrate to other employment sectors.

While legislation exists in many countries to protect the rights of employees – acts dealing with employment standards, for example, or workers' compensation – enforcement can be difficult, even more so for certain groups such as visible minorities, women, those who work part-time, and so on (England, 2008, p. 115). Enforcing one's rights through courts or tribunals is often prohibitively expensive, lengthy, and onerous, resulting in employers not having to respond to or take seriously infractions they may have committed. At almost every turn, the employer–employee relationship is distinctly weighted towards the employer.

Employees who are members of a union exercise much more power. Unions were created to protect "the weaker contracting party, balancing the interests of workers and employers" (Ojeda Avilés, 2009, p. 47). By banding together, workers as a group were able to increase their bargaining power. Yet in recent years, the strength of unions appears to be diminishing. From the rise in globalization in the past several decades to the economic meltdown of 2008 to the increased use of contract workers, labour unions have lost a great deal of influence and traction. Increased globalization (including outsourcing); the rise of contract work; the evolving nature of work, from technological advances to the decline of manufacturing; and the overall weakening of laws are but a few challenges facing labour unions today (Bronstein, 2009). Since the 2008 economic crash, many overtly political maneuvers have been launched to curtail the reach of unions. For example, the government of Wisconsin in the United States attempted in 2011 to severely restrict the bargaining rights of public sector employees. While the courts struck down much of the law in 2012, the fight is far from over, as an appeal is still likely (Greenhouse, 2012). In Ontario, Canada, the provincial government froze the wages of teachers and unilaterally banned them from striking (Howlett, 2012). The Canadian federal government has forced several unions back to work after going on legal strikes under the guise that such strikes would either damage the Canadian economy or impact public safety (see, for example, Chase et al., 2012).

Many question whether or not unions can survive in such an environment, including some unions themselves (Van Alphen, 2012). While they still represent their members and help to balance out the power employers have over employees, the number of unionized workers is on the decline.

Canada

What do courts and tribunals say about burnout?

Courts and various administrative tribunals in Canada have not given much credence to the concept of burnout. While adjudicators may mention burnout within the context of their reasons, the malady in question is eventually labelled as stress, PTSD, or another related disorder. In *Decision No. 397/92* from the Ontario Workers' Compensation Appeals Tribunal, one of the worker's psychologists labelled his disorder as "'classical burnout phenomena which manifested itself in both physical and psychological exhaustion'" (1992, para. 115).

While burnout was held by a mental health professional to be a factor in the employee's inability to work, it was essentially dismissed. The adjudicators relied on other mental health professionals, who stated "that the 'burnout' concept is not one which is included in the classification systems which psychiatrists use . . . the term 'burnout' is a rhetorical term . . . [and] that terms such as 'burn-out' . . . may be used as one approaches retirement" (sic) (1992, paras. 163–4, 167).

This statement – that burnout is simply a rhetorical term – is contrary to the research of people such as Christina Maslach (1982) who have established that burnout is indeed its own separate syndrome with an established definition and indicators (see Chapter 2 above). Ultimately, adjudicators found that the worker suffered from a psychological disability that was compensable, but labelled it as PTSD rather than burnout. In another decision, *Decision No. 1030/89*, the adjudicators provided some commentary about burnout:

> there is some agreement that the concept of burnout, when used in reference to a working situation, is based on the premise of cumulative stresses in a particular work environment. . . . In the literature, the term burnout has been described on the basis of three main features: emotional exhaustion; depersonalization toward those for whom one is responsible; and devaluation of one's achievements. However, such characterization is only one of several variations found in the literature. (1991, para. 146)

This rare recognition is promising; however, the adjudicators went on to describe why burnout is difficult to establish:

> The self-generating and self-reporting nature of burnout makes it difficult to objectively measure the actual level of discomfort experienced. . . . The overall uncertainty covering burnout as a clinical entity partly explains why burnout is not listed in major classification systems. . . . Another problem is that descriptions of burnout closely approximate descriptions of depression. (1991, para. 146)

The adjudicators ultimately decided burnout was not applicable in the case at hand.

While many have dismissed burnout as a syndrome, a handful of courts and tribunals have held burnout to be a diagnosable syndrome that, if work related, may be compensated by either employers' or workers' compensation programs. The most pertinent case in Canada is *Zorn-Smith v. Bank of Montreal.* Susanne Zorn-Smith was informed while on maternity leave that her position had been downsized. The Bank of Montreal offered her a position in another department in which she had no experience or training; as well, to remain in the position she was to complete outside courses and training in her own time. Although she felt this was not a suitable replacement position, she believed she had no other choice. On top of having no experience or training in this particular area, she was put in the position during the busiest time of year in an understaffed department where she was expected to pick up the additional workload. She was also pressured to work more on the outside courses so she would have the proper qualifications earlier.

The heavy workload and pressure from management took its toll, and less than a year after taking on this new position Zorn-Smith went out on short-term disability. She returned but the problems in the workplace were not addressed, resulting in her returning to short-term disability a year later. While her doctor considered her to be totally disabled, the Bank's medical advisor claimed she was able to return, at least on a part-time basis. Based on this information, the Bank gave Zorn-Smith an ultimatum: her short-term disability was to be cut off and she must return to work or resign. Zorn-Smith did not reply by the deadline imposed, and so the Bank terminated her employment.

Among other things, a doctor working with Zorn-Smith determined that she was indeed suffering from burnout, and the judge agreed: "The Bank was well aware that Ms. Zorn-Smith had suffered burnout in February 2000" (2004, para. 168). Zorn-Smith was psychologically disabled and was not required to respond to the ultimatum (2004, para. 74). The court held "[i]t was the responsibility of the Bank to ensure a safe workplace for its employees, a workplace that was not making them ill and unable to work" (2004, para. 103). Interestingly, the judge in this case stated that "the technical diagnosis for burnout" (2004, para. 48) is "an adjustment disorder with depressed and anxious mood" (2004, para. 48).

Only a few other cases can be found that deal with burnout, such as *Entner v. Swissair Transport Co*. One Quebec tribunal dealt with the issue of burnout in some detail in *Labrie et C.S.S.T.-Laurentides et Les Planchers 5 Étoiles Enr.* Labrie was extremely overworked for many months and ended up needing time off due to the strain of the long hours and heavy workload. The commissioner determined that the unfathomable amount of overtime and the additional responsibilities caused the worker to suffer from burnout. It was held that the employee experienced a workplace injury, diagnosed as burnout, which resulted in him being unable to perform his work duties (1999, p. 13; decision only available in French). The worker was specifically found to suffer from burnout due to workplace injury.

There are a few other Quebec tribunal decisions wherein an employee was held to be suffering from burnout. It appears tribunals in Quebec are more likely to give burnout credence than those in other parts of Canada, but it is still a struggle for the employee to come out on top in these types of cases.

What does the legislation say about burnout?

While few cases and decisions mention burnout, Canadian legislation is silent on the topic. From employment standards legislation to workers' compensation legislation, burnout is never mentioned. The Workplace Health, Safety and Compensation Commission of Newfoundland and Labrador briefly mentions burnout in policy number EN-18 on Entitlement and Mental Stress: " 'Burn out' from usual duties, workplace change, or performance demands is not compensable" (1999, p. 2). Burnout is never defined in the policy, but it is ruled out as being compensable under the *Workplace Health, Safety and Compensation Act* (1990).

The law and stress

Unlike burnout, much more commentary on the law of stress exists, although it is an area that remains in flux.

In Canada, workers' compensation schemes are a provincial responsibility. Workers employed by the federal government have their own federal statute similar in form to a workers' compensation act, the *Government Employees Compensation Act* (GECA), but if there is a claim under the act, the province in which the claimant resides administers the claim under their own workers' compensation system (Lippel and Sikka, 2010, pp. S16–S17). Federal employees' rights under GECA will be discussed in more detail below.

Currently, workers' compensation legislation in every jurisdiction allows claims for acute stress; that is, stress which quickly developed after a surprising, dangerous, or threatening event. Conversely, six jurisdictions – Manitoba, Ontario, New Brunswick, Newfoundland and

Labrador, Prince Edward Island, and Nova Scotia – specifically exclude claims based on long-term, cumulative, chronic stress (Lippel and Sikka, 2010, p. S17). For example, section 2(a) of the *Workers' Compensation Act* (1994–1995) of Nova Scotia states:

> 2 In this Act,
>
> (a) "accident" includes
>
> (i) a wilful and intentional act, not being the act of the worker claiming compensation,
>
> (ii) a chance event occasioned by a physical or natural cause, or
>
> (iii) disablement, including occupational disease, arising out of and in the course of employment,
>
> but does not include stress other than an acute reaction to a traumatic event.

This wording is similar to that of workers' compensation acts from other jurisdictions that refuse claims for chronic stress. In some jurisdictions, even more restrictions apply. This is discussed in more detail below.

British Columbia once restricted stress claims to only those that were acute stress reactions to a specific traumatic event. After *Plesner v. British Columbia (Hydro and Power Authority)* (2009) (*Plesner*), the province made changes to their act which came into effect in July 2012. The Court of Appeal in *Plesner* held that restrictions on stress claims under the act "are a violation of the equality provisions of the *Canadian Charter of Rights and Freedoms* and constitute discrimination on the basis of mental disability" (Lippel and Sikka, 2010, p. S17). The claimant had suffered a breakdown after an explosion of a natural gas pipeline; while he was not physically injured, was a sufficient distance away from the rupture, and the leak was contained in slightly over an hour, Plesner subsequently suffered from depression, PTSD, and paranoia. The Workers' Compensation Board held that his stress was chronic instead of acute, and that the incident could not be seen as traumatic due to his limited exposure to the event. On appeal to the Workers' Compensation Tribunal, it was held that his stress was an acute reaction to the event, but the Tribunal agreed with the Board that the event itself was not a traumatic event as defined in Policy 13.30 under the Rehabilitation Services and Claims Manual employed by the workers' compensation system in British Columbia.

On appeal to the Court of Appeal, the court ruled that the wording of the act, combined with Policy 13.30 (2009b), which enumerates when a worker may make a claim for mental stress, contravened the *Canadian Charter of Rights and Freedoms* (1982) (*Charter*):

> I am satisfied that the requirement of a "traumatic event" in s. 5.1(1)(a) of the Act, when read together with Policy 13.30, breaches s. 15(1) of the *Charter* by discriminating against Mr. Plesner, and other workers like him who suffer from purely mental work-related injuries, on the basis of mental disability. Workers with purely mental injuries are forced to meet a significantly higher threshold for compensation which is not required of those who suffer work-related injuries that are purely physical, or who suffer mental injuries which are linked to physical work-related injuries. (2009, para. 96)

In response to *Plesner*, British Columbia amended its legislation so that there are no exclusions to chronic mental stress claims. The new wording in the *Workers' Compensation Act* (1996) of British Columbia is the following:

> 5.1 (1)Subject to subsection (2), a worker is entitled to compensation for a mental disorder that does not result from an injury for which the worker is otherwise entitled to compensation, only if the mental disorder
>
> (a) either
>
> (i) is a reaction to one or more traumatic events arising out of and in the course of the worker's employment, or
>
> (ii) is predominantly caused by a significant work-related stressor, including bullying or harassment, or a cumulative series of significant work-related stressors, arising out of and in the course of the worker's employment.

British Columbia joins six other jurisdictions – those being Quebec, Saskatchewan, Alberta, Northwest Territories, Nunavut, and the Yukon – none of which have exclusions to chronic mental stress claims in their workers' compensation legislation (Lippel and Sikka, 2010, pp. S17–S18).

While these jurisdictions expand the ability of workers to make mental stress claims, some reiterate that employees must still meet a higher threshold for a mental injury than for a physical injury. In reference to the British Columbia legislation, some commentators point out that

> "[f]or any other physical injury, the work cause need only be 'significant,' and if there were no disabilities due to the injury before the work accident/exposure, work is considered to be the cause. For psychological injury, however, the work related cause must be 'predominant,' which means the board can delve into the worker's private life and dig up any other stressors to blame for the condition, regardless of whether those issues were in any way disabling prior to the workplace stressors. This is just plain discriminatory."
>
> [Union President Michelle] Laurie said the requirement that a mental condition be diagnosed by a psychiatrist or psychologist, rather than a family doctor as it has worked in the past, is problematic.
>
> "It takes a long time to get an appointment with a psychiatrist, even in urban centres, and even longer if the injured worker lives outside of the Lower Mainland. People should be outraged about how workers with psychological injuries are being treated." (Sandborn, 2012)

Lippel and Sikka also submit that in the seven jurisdictions that allow for claims involving chronic stress, it has been held by the courts and tribunals that the precipitating stressor must be unusual or abnormal; that is, outside normal everyday stressors. What the criteria are for unusual or abnormal is not clear (2010, pp. S17–S18).

In several jurisdictions, whether allowing claims for chronic mental stress or not, decisions dealing with management of workers' employment, such as disciplinary matters or termination, are not subject to stress claims. Additionally, in either the legislation or the policies adopted by some Workers' Compensation Departments in many jurisdictions (for example, British Columbia), only those mental disorders as defined in the *Diagnostic and Statistical Manual of Mental Disorders*, or some other similar diagnostic manual, may be the subject of a mental disorder claim.

These extra criteria further restrict the ability to file a claim, making it quite difficult for mental stress-related claims to be successful.

Despite such restrictions, a chronic pain case from the Supreme Court of Canada has now opened the door on the national level on conditions normally excluded from workers' compensation acts. In *Nova Scotia (Workers' Compensation Board) v Martin; Nova Scotia (Workers' Compensation Board) v Laseur* (two cases heard together), two claimants were suffering chronic pain due to workplace accidents; however, in the Nova Scotia *Workers' Compensation Act* and related regulations, chronic pain was specifically excluded as being covered. The Supreme Court of Canada found this exclusion to be in violation of the *Charter*, as those experiencing chronic pain were provided different treatment based on the nature of their disability (2003, para. 5). The Industrial Accident Victims' Group of Ontario and Community Legal Education Ontario co-published a manual on workers' compensation issues wherein the authors argue that this case applies to workers who have suffered some sort of mental disability or injury on the job: "the . . . exclusion from benefits of workers based on their type of psychological disability is discriminatory, and not justifiable under the *Charter*" (2006, ch. 11, pp. 11–15). Such *Charter* (1982) arguments will likely be more common in the future.

Federal employees have their own compensation act, as mentioned above – GECA. The province in which the federal worker lives and works will adjudicate the claim on behalf of the federal government. There has been an issue, though, of some adjudicators transposing the language of their own provincial workers' compensation legislation into GECA. Some argue that exclusions found in provincial acts have no merit when adjudicating a claim: "it has been held on several occasions that [federal] employees . . . will not be subject to the statutory exclusions in provincial legislation, given that the federal Act provides for no such exclusion" (Lippel and Sikka, 2010, p. S17). According to this argument, federal employees have broader access to workers' compensation claims due to the looser language of the federal act, including mental health claims.

In a case from Newfoundland and Labrador, this distinction was made slightly clearer. In *Canada (Royal Canadian Mounted Police) v. Rees*, a jail guard employed by the Royal Canadian Mounted Police (RCMP) was required to provide comments on the actions of his supervisor. Rees believed his comments would be kept confidential, but they were in fact shown to his supervisor. The supervisor then began a campaign of harassment and humiliation to the point where Rees became stressed and eventually took sick leave. Rees eventually filed a statement of claim alleging harassment, breach of duty of care, and negligence. A question of jurisdiction to hear the case arose with the RCMP claiming Rees had to go through the workers' compensation system instead of a civil action. In the process of finding that Rees was able to file a civil claim, the court held that the language in GECA must be used rather than that of the workers' compensation legislation in the province in which the claim was filed:

> The fact that language comparable to that in provincial legislation, which specifically excludes gradual onset stress, is not employed

> under the federal Act further supports the conclusion that gradual onset stress falls within the scope of "accident" for which compensation may be payable under the Government Employees Compensation Act. (2005, para. 32)

Other courts agree that the wording in GECA cannot be limited by stricter wording in a provincial workers' compensation statute, although there have been decisions that have found otherwise, such as *Re Morrison Estate*, in which the Nova Scotia Court of Appeal held

> the language in s. 4(1) of GECA is sufficiently broad and inclusive to embrace the language creating entitlement in the Nova Scotia Act; thus a worker entitled under GECA to claim compensation in Nova Scotia is by necessary implication also entitled to the benefits, and is bound by any restrictions contained in the provisions of the Nova Scotia Workers' Compensation Act. (2003, para. 45)

While there seems to be division on this issue, some provincial Workers' Compensation Boards are adopting policies that make the interplay between GECA and the provincial workers' compensation statute a bit clearer. Subsequent to the *Re Morrison Estate* case, the Workers' Compensation Board of Nova Scotia determined that federal employees filing chronic stress claims under GECA can be compensated as long as the criteria outlined in their GECA policy is met (2005, Policy 1.3.6; WorkSafeBC, 2009a, p. 15). WorkSafeNB, the workers' compensation commission in New Brunswick, states in Policy 21–011 that "A claim for mental stress may be compensable if it meets the criteria outlined in section 4 of *GECA*, which WorkSafeNB uses to determine if an accident arose out of and in the course of employment" (2001, s. 3.1, p. 7).

It is interesting to note that a number of decisions and cases in the 1990s in some jurisdictions did allow claims for chronic stress, leading to amending statutes so that they explicitly eliminate chronic stress as an eligible claim (WorkSafeBC, 1998; Shortt, 1995). In *Dowling v. Prince Edward Island (Workers' Compensation Board)* (1994), the Prince Edward Island Court of Appeal held that there was nothing in the definition of accident or employment under the then version of the *Workers' Compensation Act* that would not allow a claim for chronic stress. Subsequent to this decision, the government of Prince Edward Island amended the statute so that chronic stress was specifically eliminated as an eligible claim (WorkSafeBC, 1998, p. 7). Similarly, the Ontario workers' compensation statute was amended in 1998 so that chronic stress was no longer eligible for a claim (Gunderson and Hyatt, 2000, p. 358). Other jurisdictions followed suit (WorkSafeBC, 1998, p. 18; Lippel and Sikka, 2010, p. S17).

Even after such changes, "case law shows a surprising variation in the ways in which these [exclusionary] provisions have been interpreted and applied" (Lippel and Sikka, 2010, p. S17). In a Nova Scotia Court of Appeal case, a worker was chastised by his supervisor, resulting in his suffering from PTSD. The employer argued that the event would not be seen as traumatic by a reasonable person, but the court held that this was a traumatic event for the worker (who had suffered from past anxiety), and was not a common, usual experience (Lippel and Sikka, 2010, p. S17; *Children's Aid Society of Cape Breton-Victoria v. Nova Scotia (Workers' Compensation Appeals Tribunal)* (2005)).

Stress at work has recently been given more attention due to a report about stress in the workplace published in 2009 by the Mental Health Commission of Canada. Martin Shain, the author of the report, found that there is a great "deal of uncertainty concerning where the boundaries of liability for mental injury at work lie. This situation is due in part to the rapidly evolving nature of legal developments in this area" (p. 5). In a 2010 follow-up, Shain stated that "[a] perfect legal storm is brewing in the area of mental health protection at work. This storm brings with it a rising tide of liability for employers in connection with failure to provide or maintain a psychologically safe workplace" (Shain, 2010, p. 24). Mental injury in the workplace is an increasingly important issue that must be collectively addressed. A lack of attention is more likely to lead to strife, grief, and economic distress for employees and employers alike.

While we have focused mainly on workers' compensation schemes, other areas of law such as human rights, occupational health and safety, and labour law are also important to note, and each will be discussed in brief below.

Human rights, as Shain notes in his 2009 report, are based on the idea of equal rights for all, which is buoyed by the *Charter* (p. 61). There is a human rights act in every jurisdiction in Canada, all enumerating grounds on which discrimination is prohibited, one of which is disability. Among other issues that arise when discussing human rights law, two in particular are important to note: 1) "[e]mployers are responsible for the maintenance of workplace environments that are free of, and not poisoned by discriminatory policies and practices" (Shain, 2009, p. 62), and 2) there is a duty to accommodate an employee who has limitations due to disability (Shain, 2009, p. 67). More and more human rights cases are being decided based on issues of mental illness, disability, and injury.

The goal of occupational health and safety legislation is to ensure the workplace is a safe, healthy place for employees. The prevention of injuries and disease is the primary focus of such legislation (Neumann and Sack, 2012, s. 5.2.2.1). To date, most such legislation does not mention or include mental health specifically, other than Saskatchewan (Shain, 2009, p. 70); there seems to be an emphasis on harassment in the workplace as the key form of psychological injury or illness (Shain, 2009, p. 74). As awareness of mental health issues in the workplace improves, it is likely the occupational health and safety schemes in each jurisdiction will have to broaden their language to be more inclusive of unhealthy or unsafe conditions that impact mental health.

In terms of labour law, as Shain states in his 2010 report, "[l]abour [l]aw offers employees covered by collective agreements a strong shield against mental injuries in so far as it allows them access to contractual remedies *while the employment relationship is still intact*" (p. 15; italic in original).

Members of a union have their own grievance scheme to assist them with complaints about harassment, discrimination, and so forth. They also can rely on provincial labour relations legislation, occupational health and safety acts, and human rights codes (Shain, 2009, p. 54; Shain, 2010, p. 15).

Along with numerous agreements and legislation to bolster claims, employees covered by a collective agreement are able to rely on their union to assist them through the grievance process. Such support may be crucial; when suffering a mental injury, it is often very difficult to act effectively as one's own advocate. As well, the grievance process itself can be quite draining and onerous, so having a union to assist is extremely beneficial in lessening the overall stressful impact.

United States

What do courts and tribunals say about burnout?

Several courts and tribunals in the United States have dealt with the issue of stress on the job, but relatively few have dealt with burnout. As in Canada, burnout is often referred to as stress, PTSD, or other related disorders.

Some courts and tribunals have found that mental disorders such as stress and burnout are not compensable. In *Andolsek v. City of Kirtland*, for example, the plaintiff police officer filed a claim with the worker's compensation board asserting that he was suffering from burnout, PTSD, and depression related to working on a particularly difficult case dealing with cult killings (1994, p. 335). His claim was dismissed by several tribunals and review boards, as well as at the trial level. These adjudicators disallowed his claim because it "alleged a psychiatric condition without physical trauma" (1994, p. 335). The Court of Appeal agreed with the lower court ruling and held that

> [p]ursuant to Ohio's workers' compensation statutory scheme, compensation is sanctioned for a worker who is disabled by a mental condition resulting from a compensable work-related physical injury ... compensation is not available for workers suffering mental conditions caused solely by job-related stress. (1994, pp. 335–6)

Some courts, however, have recognized burnout as a legitimate ailment. In *In re Complaint as to the Conduct of Robert J. Loew* (1982), a lawyer was brought before a disciplinary committee for failing to follow the Code of Professional Responsibility. The lawyer had delayed working on a client file, failed to file an appeal brief, rarely answered the client's calls, misled the client about filing documents with the court, and failed to relinquish the file to the client when requested (1982, pp. 808–10). A psychiatrist eventually diagnosed the lawyer as suffering from burnout. While the court does not mention the word itself, the testimony from the psychiatrist was completely on burnout. In the end, the court held "the conduct of the accused was an isolated event caused by emotional difficulties with which he is now dealing effectively" (1982, p. 812). Instead of disbarment, the lawyer was suspended for a minimum of 30 days, and could only resume practice if a psychiatrist provided a written statement stating that the lawyer has dealt with his emotional problems (1982, p. 812). Justice Peterson held that this case should be "required reading for every lawyer, for almost every practicing lawyer becomes involved in situations which create pressures and stresses akin to those which are present in this case" (1982, p. 813). Clearly, burnout in lawyers is not considered unusual by the courts. In fact, two other cases involving the same lawyer were brought to court, but as the incidents in question took place when the lawyer was experiencing the same burnout episode, the courts found that no further sanctions were needed (*In re Complaint as to the Conduct of Robert J. Loew*, 1983, p. 681; *In re Complaint as to the Conduct of Robert J. Loew*, 1984, p. 337).

In a similar case, *State ex rel. Oklahoma Bar Ass'n v. Schraeder*, the judge agreed that the lawyer was experiencing burnout when the actions for which he was brought before a disciplinary committee took place, yet the judge also stated in no uncertain terms that "emotional or psychological disability, though it may serve to reduce the actor's ethical culpability, does not immunize one from imposition of disciplinary measures that are necessary to protect the public" (2002, p. 580, emphasis removed). In this case, the lawyer was suspended from practice for 30 days.

In *Reidy v. Travelers Ins. Co.* (1996), the plaintiff claimed his employer discriminated against him based on age and disability. The judge found that the plaintiff was handicapped after being diagnosed with permanent burnout, yet also found that the plaintiff had not proven he was discriminated against due to this handicap. The lower court's ruling was affirmed on appeal (*Reidy v. Travelers Ins. Co.* (1997)).

These cases tell us that some courts and tribunals have recognized burnout as a disorder; nonetheless, it is still rare for a claimant to be successful.

What does the legislation say about burnout?

No federal statutes or regulations, or any state statutes deal specifically with workplace burnout. Some state regulations mention workplace or career burnout, but such instances are usually in general terms with no applicable definition or context. In Idaho, Rule 24.13.01, Rules Governing the Physical Therapy Licensure Board, states at section 250.06.b.v.(4).viii that physical therapists will be given continuing professional credit for "[p]articipating in courses that have personal skills topics: career burnout, communication skills, human relations, and other like topics" (2012). Similarly, in West Virginia, Rule 64-85, Alzheimer's/Dementia Special Care Units and Programs at section 4.1.c.10 states that staff members dealing with Alzheimer's and dementia patients must complete training that includes "[s]taff burnout prevention."

As Jauregui and Schnall reiterate, "a number of health conditions, such as burnout . . . which appear to be the result of exposure to workplace psychosocial stressors, are not recognized as yet as work-related disorders" (2009, pp. 153–6). The authors proffer that there is a reluctance to include disorders such as burnout as workplace injuries as it might overburden the workers' compensation system, and there is still the strongly held belief that such disorders are mainly the result of individual differences (2009, p. 156). Such beliefs must be overcome before burnout can truly be recognized as a work-related disorder.

The law and stress

Some scholars and psychologists recognize that "[c]hronic exposure to psychosocial stressors in the work environment . . . can be as noxious to your health as physical demands and chemical toxins" (Dobson and Schnall, 2009, p. 114). In the workers' compensation system, however, the impact of work on stress has not yet been consistently accepted as compensable.

There are typically three categories of stress claims in workers' compensation systems: physical-mental, mental-physical, and mental-mental. Physical-mental stress claims are those where a physical injury appears first and leads to a mental disorder (deCarteret, 1994, p. 495). An example of this type would be breaking a leg on the job that did not heal properly, resulting in chronic pain. If a person suffers depression and anxiety related to the chronic pain, his or her mental disorder came about as a result of a physical injury.

A mental-physical claim is when an employee suffers a mental health injury that then results in physical injury or illness (deCarteret, 1994, p. 495). For example, if a worker suffers a heart attack due to a stressful work environment, the physical injury was a result of a mental stimulus. Finally, a mental–mental claim is when some sort of mental distress results in a mental injury or illness (deCarteret, 1994, p. 495). An example would be if a worker suffers harassment at work and then develops a panic disorder (Meriano, 2001, p. 258).

The mental-mental claim is the hardest of the three to prove. Each claim encompasses differing elements of proof, with the mental-mental claim having the most difficult burden. For physical-mental, Meriano explains the elements that must be proved: that there is a mental injury, this mental injury is a result of a work-related physical injury, and there is a causal connection between the two (2001, p. 259). Under the mental-physical injury category, the elements of proof include establishing that the injury happened as a result of the employment, and that there was a causal connection between the mental and physical injury (Meriano, 2001, p. 258).

Proving a mental-mental injury "is the hardest . . . because it is difficult to present objective data regarding mental injury" (Meriano, 2001, p. 259). Often one must prove a mental injury exists through reliance on medical experts. One must also prove that the employee "was exposed to unusual and extraordinary conditions in [one's] employment . . . and . . . that these unusual and extraordinary conditions were the proximate cause of [one's] mental breakdown" (Meriano, 2001, p. 259).

Having to prove that conditions were unusual and extraordinary can be extremely demanding. What one person may consider unusual another may consider a standard condition of the work environment. In some jurisdictions, the conditions must also be sudden and unexpected.

As in Canada, the workers' compensation system in the United States is not a federal matter; each individual state designs their system differently according to their separate laws and policies. While some states allow claims for mental injury, others will not. But as Ivancevich et al. state, "[i]n the case of a psychologically related injury, the courts are reluctant to compensate the worker because it is difficult to establish either the mechanism or the extent of the injury" (1985, p. 62). That being said, they argue that courts are more apt to allow for compensation for mental injury if said injury is a result of a "discrete, identifiable accident" (1985, p. 62) rather than a build-up of stress over time that did not arise out of one particular incident (1985, p. 62). This does not mean, however, that the latter is never successful. *Carter v. General Motors* (1960) (as discussed in Ivancevich et al.) involved an assembly-line worker whose supervisor criticized and ridiculed him for his difficulty in keeping up with the line. Eventually, the worker, who had evidenced some mental illness in the past, broke down under the strain. The court found that there was no one single incident that led to the breakdown, but rather a series of events taking place over a long period of time (Ivancevich et al., 1985, p. 62). These types of incidents – where there is chronic, long-term stress unassociated with any one particular distressing incident – appear to be the basis of many of the stress-related illness cases now before the courts (Ivancevich et al. 1985, p. 66).

Since the workers' compensation system varies by state, each state has arrived at its own way of viewing mental-mental claims, but there are similarities:

(1) In some states, mental-mental claims have been held to be compensable if the source of the stress is over and above what one would experience in day-to-day life (i.e. it must be unusual or abnormal) (deCarteret, 1994, p. 496). See, for example, Maine and Michigan (Tucker, 2010, p. 471).

(2) In other states, the source of stress does not have to be over and above what one would generally experience in day-to-day life

(deCarteret, 1994, p. 496). See, for example, Kentucky and Utah (Smith, 1995, pp. 230–1).

(3) In some states, a claim for mental-mental injury can only be compensated if the injury was a result of an abrupt and severe event (deCarteret, 1994, p. 496). See, for example, Louisiana (Tucker, 2010, p. 469).

(4) Some states will not compensate for stress claims that developed gradually unless there is a physical manifestation (deCarteret, 2010, p. 496). See, for example, Montana and Florida (Tucker, 2010, p. 469).

(5) The courts in a few states (for example, Delaware) have held that an injury may include a mental injury (Tucker, 2010, pp. 468–69), whereas in states such as Minnesota, mental injury is not included (Tucker, 2010, p. 469).

(6) In some states, "benefits for mental-mental claims [are allowed] so long as the mental impairment is causally related to the stress" (Tucker, 2010, p. 475). See, for example, Alaska and California (Tucker, 2010, p. 475).

One of the first cases in the United States to deal with a work-related mental injury was *Bailey v. American General* (1955, as discussed in Ivancevich et al.). Two ironworkers were on a high scaffold when it collapsed; one was saved by his cable, but the other fell to his death. The survivor was traumatized – after he returned to work, he would black out, was paralyzed when on the scaffold, had trouble sleeping, and was extremely sensitive to pain. "The court found that he was disabled, that the disability was work related, and that the employer owed him compensation" (Ivancevich et al. 1985, p. 62). Several took this case as a turning point, proving that mental injuries as a result of a workplace incident could be compensable. If one reviews the judgment, however, it shows that while the court did agree that Bailey suffered a mental injury, it was really compensating him for his physical injuries: "[w]hile liberal for its day, this decision, used as a precedent, subsequently denied Texas workers compensation for most stress-related injuries because it based the award on resulting physical rather than psychological illness" (Ivancevich et al., 1985, p. 62). Even though the mental injury was recognized, the worker only received compensation due to the subsequent physical injuries.

There were several changes to workers' compensation legislation in the 1990s as a result of an increase of stress-related claims. In California, there was an increase of 700 percent between 1981 and 1991 (deCarteret,

1994, p. 497). In the early 1990s, a worker only had to show that the workplace was 10 percent responsible for the psychological injury (deCarteret, 1994, p. 497). This resulted in a huge number of claims, some of which are rather suspect. For example, a doctor who had been found guilty of "fraudulently billing insurance companies" (deCarteret, 1994, p. 497) for millions of dollars claimed that he suffered extreme stress while awaiting his sentence, and was subsequently awarded workers' compensation benefits (deCarteret, 1994, p. 497). A restaurant in California had shut down "and 115 of its 119 employees filed stress claims. Attorneys had stationed themselves outside the door the day of the lay-offs to intercept the newly unemployed" (deCarteret, 1994, p. 497). Due to this influx of stress-related claims, the laws in California were changed so that work-related stress now "must be *the* predominant cause of the injury" (DeFrank and Ivancevich, 1998, p. 59; italic in original).

In addition to workers' compensation claims, two federal acts come into play: the *Occupational Safety and Health Act* (OSHA) and the *Americans with Disabilities Act* (ADA). Other federal acts, such as the *Federal Employers Liability Act*, deal with compensation for workers in specific industries, in this case railway workers. Such acts particular to a specific employment sector will not be discussed in this chapter.

OSHA, enacted in 1970, presented a framework for providing a safe work environment (Harthill, 2010, pp. 1264–5). The act does not specifically include mental illnesses as a work injury; even though some courts have been reluctant to determine that OSHA deals with mental injuries (Abu Al Rub, 2000, p. 133), some argue that OSHA does indeed deal with this issue, as long as the political will is there:

> While OSHA has not yet systematically addressed the problem of psychological problems in the workplace, OSHA's legislative mandate would permit them to tackle this issue through new regulations. . . . Some scholars have suggested that the lack of psychological regulation is due to the agency's prioritizing of more traditional, physical health issues and the "perceived exigency" of those problems. (Dillard, 2008, p. 403)

The ADA was enacted in 1990 to provide protections for those who have disabilities. The ADA maintains a structure different from workers' compensation schemes, compelling employers "to provide equal opportunities for individuals with disabilities concerning employment" (Abu Al Rub, 2000, p. 133). While the ADA does cover impairments that limit how people function, it does not cover all impairments, with some courts ruling that stress does not qualify (Abu Al Rub, 2000, p. 133). While few cases have dealt with stress under the ADA, some argue that ADA could indeed apply in such cases:

> courts should ... [look] to the manifestations rather than the etiology of disability in determining whether the plaintiff is disabled under the ADA. People *can* drive each other crazy; hostile, abusive treatment *can* trigger underlying vulnerabilities; extraordinary stress can be the final straw that breaks the worker's back. (Stefan, 1998, p. 844; italic in original)

Undoubtedly, when it comes to stress and burnout, the law lags far behind the realities of the modern workplace.

United Kingdom

What do courts and tribunals say about burnout?

As with Canada and the United States, there are few specific cases and tribunal decisions dealing with burnout in the United Kingdom, although several do deal with stress. In *Stafford and Rural Homes Ltd v. Hughes*, for example, Hughes began suffering from a mental disorder, yet rather than assist him with counselling or other support, his employer sent him a disciplinary letter, leading to Hughes' complete breakdown. At the tribunal level, it was held "that the Claimant had suffered 'burnout' and that the pressures upon him had reached a point where he was unable to tolerate any more pressure" (2009, para. 69). The Employment Appeal Tribunal agreed with the tribunal decision and found that the employer made no reasonable attempts to assist the claimant or make reasonable adjustments.

While a few other cases and tribunal decisions mention burnout as an issue, there has to date been little to no substantive discussion. This is also true of the case law from the Court of Justice of the European Union and the European Court of Human Rights, two extra-territorial courts which exercise some jurisdiction over the UK. Neither side has yet to decide any cases involving burnout in the UK workplace.

What does the legislation say about burnout?

No legislative provisions in the United Kingdom specifically deal with burnout, nor does any legislation emanating from the European Union (EU). However, legislation from both the UK and EU deals with the matter of workplace stress.

The law and stress

As a member state of the EU, the UK must follow and implement EU directives. Directive 89/391/EEC, the EU Health and Safety Framework Directive, mandates that all states must "ensure the safety and health of workers in every aspect related to the work" (1989, art. 5.1). Further, the directive explicitly states that the prevention and elimination of occupational risk is paramount (1989, art. 1.1). Article 6.2.d sets out that employers ought to "adapt . . . the work to the individual" (1989) (while burnout, stress, PTSD, and other stress-related disorders are not mentioned specifically, as will be shown below, it has been held that such stress-related disorders are workplace risks). The UK executed this directive by implementing various legislation, such as *The Management of Health and Safety at Work Regulations 1992; The Management of Health and Safety at Work Regulations 1999; The Health and Safety (Consultation with Employees) Regulations 1996*; and *The Workplace (Health, Safety and Welfare) Regulations 1992.*

Legislators in the UK recognized the problem of stress in the workplace going back to 1974. In various sections throughout the *Health and Safety at Work etc. Act 1974*, it is stated that a mental condition or illness may be considered damage or personal injury. For example, the definition of " 'personal injury' includes any disease and any impairment of a person's physical or mental condition" (1974, s. 53(1)).

As such, there is some recognition in the UK of the dangers of workplace stress. But legislation is only one element; a serious question remains as to how courts and tribunals are dealing with the issue.

One groundbreaking case on workplace stress is *Walker v. Northumberland County Council*. John Walker was a social services officer with the county. He supervised other employees, had an extremely heavy workload, and worked in an area that had an exceedingly high number of difficult cases involving abuse. After a number of years, he experienced a mental breakdown and was off work for several months. Walker wanted to return, but his doctor advised that the only way he could return was if his responsibilities were drastically reduced. Walker spoke to his supervisor, who agreed to make some changes and provide assistance, but ultimately did not agree to split the large district Walker had been handling into two separate districts. When Walker returned to work, only limited temporary assistance was made available; after a few weeks the assistance was removed and Walker was told to ask if he needed more help. Additionally, the workload was not diminished but was in fact heavier, as Walker had to deal with the paperwork that had accumulated during the time he was off sick. Walker suffered another breakdown six months after his return, and had to cease work permanently. His employer dismissed him due to his ill health, and Walker sued for a breach of duty of care.

The court held that a second breakdown was foreseeable since Walker was essentially provided no assistance and no reduction in work. The judge noted,

> [t]here has been little judicial authority on the extent to which an employer owes to his employees a duty not to cause them psychiatric damage by the volume or character of the work which the employees are required to perform. It is clear law that an employer has a duty to provide his employee with a reasonably safe system of work and to take reasonable steps to protect him from risks which are reasonably foreseeable. Whereas the law on the extent of this duty has developed almost exclusively in cases involving physical injury to the employer as distinct from injury to his mental health, there is no logical reason why risk of psychiatric damage should be excluded from the scope of an employer's duty of care or from the co-extensive implied term in the contract of employment. (1994, p. 749)

The court recognized that psychiatric illnesses ought to be included with physical illnesses when contemplating and discussing workplace health and safety; however, the judge only dealt with the second breakdown, holding that the first was not something that could have been forecast by the managers, as there was no indication beforehand that Walker could not handle the work. Hence, it was held that the employer did not breach its duty of care to Walker prior to his first breakdown. After Walker returned to work with no assistance and no reduction in work, it was reasonably foreseeable that he would suffer another breakdown. The negligence in not changing the work situation of Walker led to a breach of duty of care, and the employer was liable for Walker's second breakdown.

Since this case, other cases dealing with workplace stress have been decided. In *Armstrong v. Secretary of State for the Home Department*, for example, the judge followed the principles identified in *Walker* and outlined a three-pronged analysis to determine whether or not psychiatric injury was reasonably foreseeable on the part of the employer:

i) Did the work which the claimant was required to do create a reasonably foreseeable risk of psychiatric injury?
ii) If so, was the system of work in place a reasonable one?
iii) If there was a reasonably foreseeable risk of psychiatric injury and the system of work was not reasonable, did the failure to adopt a reasonable system cause the injury of which the claimant complains? (2001, para. 17)

While it was ultimately determined that the employer in *Armstrong* could not have reasonably foreseen the psychiatric injury, resulting in the claim being dismissed, the principles as founded in *Walker* were reaffirmed.

In 2002, the England and Wales Court of Appeal (Civil Division) released the decision of *Hatton v. Sutherland and other Appeals*, which joined together four appeals from four separate judgments. Each appeal was from employers who had been found liable for their employees' psychiatric illness resulting from workplace stress. The Court of Appeal allowed the appeals in three cases, dismissed the appeal in one. In the cases where the appeals were allowed (the plaintiffs of Hatton, Barber, and Bishop), it was held that it was not reasonably foreseeable that the employees would suffer a psychiatric illness due to work-related stress. In the case of plaintiff Jones, however, it was found that the psychiatric illness she suffered was reasonably foreseeable:

> there was evidence before the judge which entitled him to reach the factual conclusions he did, and from those to conclude that the indications of risk to mental health were strong enough for a reasonable employer to think that he should do something about it, not least because senior management did think that there was something they should do. (2001, para. 210)

At paragraph 43, the judges established a thorough 16-point assessment for determining liability for stress in the workplace:

From the above discussion, the following practical propositions emerge:

(1) There are no special control mechanisms applying to claims for psychiatric (or physical) illness or injury arising from the stress of doing the work the employee is required to do. . . . The ordinary principles of employer's liability apply. . . .

(2) The threshold question is whether this kind of harm to this particular employee was reasonably foreseeable . . . this has two components (a) an injury to health (as distinct from occupational stress) which (b) is attributable to stress at work (as distinct from other factors). . . .

(3) Foreseeability depends upon what the employer knows (or ought reasonably to know) about the individual employee. Because of the nature of mental disorder, it is harder to foresee than physical injury, but may be easier to foresee in a known individual than in the population at large. . . . An employer is usually entitled to assume that the employee can withstand the normal pressures of the job unless he knows of some particular problem or vulnerability. . . .

(4) The test is the same whatever the employment: there are no occupations which should be regarded as intrinsically dangerous to mental health. . . .

(5) Factors likely to be relevant in answering the threshold question include:

   (a) The nature and extent of the work done by the employee. . . . Is the workload much more than is normal for the particular job? Is the work particularly intellectually or emotionally demanding for this employee? Are demands being made of this employee unreasonable when compared with the demands made of others in the same or comparable jobs? Or are there signs that others doing this job are suffering harmful levels of stress? Is there an abnormal level of sickness or absenteeism in the same job or the same department?

   (b) Signs from the employee of impending harm to health. . . . Has he a particular problem or vulnerability? Has he already suffered from illness attributable to stress at work? Have there recently been frequent or prolonged absences which are uncharacteristic of him? Is there reason to think that these are attributable to stress at work, for example because of complaints or warnings from him or others?

(6) The employer is generally entitled to take what he is told by his employee at face value, unless he has good reason to think to the contrary. He does not generally have to make searching inquiries of the employee or seek permission to make further inquiries of his medical advisers. . . .

(7) To trigger a duty to take steps, the indications of impending harm to health arising from stress at work must be plain enough for any reasonable employer to realise that he should do something about it. . . .

(8) The employer is only in breach of duty if he has failed to take the steps which are reasonable in the circumstances, bearing in mind the magnitude of the risk of harm occurring, the gravity of the harm which may occur, the costs and practicability of preventing it, and the justifications for running the risk. . . .

(9) The size and scope of the employer's operation, its resources and the demands it faces are relevant in deciding what is reasonable; these include the interests of other employees and the need to treat them fairly, for example, in any redistribution of duties. . . .

(10) An employer can only reasonably be expected to take steps which are likely to do some good: the court is likely to need expert evidence on this. . . .

(11) An employer who offers a confidential advice service, with referral to appropriate counselling or treatment services, is unlikely to be found in breach of duty. . . .

(12) If the only reasonable and effective step would have been to dismiss or demote the employee, the employer will not be in breach of duty in allowing a willing employee to continue in the job. . . .

(13) In all cases, therefore, it is necessary to identify the steps which the employer both could and should have taken before finding him in breach of his duty of care. . . .

(14) The claimant must show that that breach of duty has caused or materially contributed to the harm suffered. It is not enough to show that occupational stress has caused the harm. . . .

(15) Where the harm suffered has more than one cause, the employer should only pay for that proportion of the harm suffered which is attributable to his wrongdoing, unless the harm is truly indivisible. It is for the defendant to raise the question of apportionment. . . .

(16) The assessment of damages will take account of any pre-existing disorder or vulnerability and of the chance that the claimant would have succumbed to a stress-related disorder in any event. (2002)

After studying these principles, it is clear employers have a duty of care to prevent psychiatric injuries, yet the enumerating circumstances listed may lead one to hold that the test favours the employer. That said, it is certainly not impossible for claimants to win their case, as we saw with the Jones claimant in the *Hatton* case.

As further proof, one of the losing claimants appealed further to the House of Lords. In *Barber v. Somerset County Council*, the court overturned the appeal court's decision and reinstated the trial court decision. It found that the claimant's psychiatric illness due to stress in the workplace was reasonably foreseeable by the employer. Nonetheless, even the House of Lords admits that the case was not at all straightforward:

> My Lords, the issue of breach of the council's duty of care to Mr Barber was in my view fairly close to the borderline. It was not a clear case of a flagrant breach of duty any more than it was an obviously hopeless claim. But the judge, who saw and heard the witnesses . . . came to the conclusion that the employer was in breach of duty, and in my view there was insufficient reason for the Court of Appeal to set aside his finding. The Court of Appeal was concerned about the timing of the breach, but for my part I do not think there is much room for doubt about that. The employer's duty to take some action arose in June and July 1996, when Mr Barber saw separately each member of the school's senior management team. It continued so long as nothing was done to help Mr Barber. The Court of Appeal evidently considered that Mr Barber was insufficiently forceful in what he said at these interviews, and that he should have described his troubles and his symptoms in much more detail. But he was already suffering from depression . . . What the Court of Appeal failed to give adequate weight to was the fact that Mr Barber, an experienced and conscientious teacher, had been off

> work for three weeks (not two weeks, as the Court of Appeal thought) with no physical ailment or injury. His absence was certified by his doctor to be due to stress and depression. The senior management team should have made inquiries about his problems and seen what they could do to ease them, in consultation with officials at the council's education department, instead of brushing him off unsympathetically . . . or sympathising but simply telling him to prioritise his work. (2004, para. 67)

The employer ought to have taken some action once the claimant's illness became apparent, and its lack of action led to it being found liable. Once an employer becomes aware of a potential stress-related problem, it is legally required to make changes in the work environment so as to reduce stress and adequately support the worker (Cox, 2004, p. 36).

While some have been successful in their stress claims, it is an uphill battle with many claimants failing to prove their case. Even in a jurisdiction where both legislation and the courts have recognized that stress may result in workplace injury, it is difficult for claimants to have their employers held liable.

As Barrett has stated, "it is apparent that the victim of stress faces no easy task if intending to claim compensation from an employer for psychiatric stress" (2004, p. 348). Barrett does hold out hope that an employer's duties regarding psychiatric injury will become clearer once the Health and Safety Executive (HSE) releases its management standards (2004, p. 349). The HSE is an arm's length non-departmental government body created by the *Health and Safety at Work etc. Act 1974*, and one of its duties is to enforce the act (HSE, Enforcing Health and Safety). It also drafts various documents and standards. While the standards are not compulsory, by following them organizations "will normally be doing enough to comply with the law" (HSE, 2007, p. 2). The Management Standards for work-related stress outline the main sources of stress at work (demands, control, support, relationships, role, change) and suggest ways organizations can overcome such conditions (HSE, What are the Management Standards?). These standards have been in place for only a few years, but their implementation has been a challenge. Mellor et al. found that barriers to implementation include organizational change and upheaval, weak management support, a dearth of expertise, inadequate resources or lack of resources, lack of guidance, and outside interference (2011, pp. 1042–4). While the standards are a good start, much more needs to be done.

Conclusion

Burnout and stress are real problems that negatively affect everyone touched by them. Why, then, when there appears to be so much evidence on the prevalence of such illnesses in society, does it seem employers, governments, courts, and tribunals are reluctant to recognize it? Along with the fear of increased workers' compensation claims or damages that may be awarded in a civil action, the changing nature of work is likely a significant factor:

> The emergence of mental injury as a legal cause of action . . . is not an isolated occurrence but rather an expression of a profound and progressive evolution of the employment relationship itself.
>
> Changes in attitudes and beliefs concerning the responsibilities of employers with regard to the protection of employee mental health, which have taken 150 years to evolve, are not likely to be easily turned back or reversed. (Shain et al. 2011, p. 22)

The nature of work in the western world has changed dramatically over the last 50+ years. Where people once worked on farms or in manufacturing, the majority now work in offices or cubicles for 8+ hours per day, often surrounded by other people for the majority of the time. The pressures of employment have changed – from working on a fast-moving manufacturing line or bringing in the crops (in both instances using loud, hazardous equipment), to sitting for hours a day in front of a computer working in close contact with co-workers and supervisors. While there still exist physical strains in office jobs (e.g., carpel tunnel syndrome), there is an increasing amount of mental strain as well. Working with others can be a wonderful experience and can make the job enjoyable, but if other workers or supervisors make the day-to-day work miserable or stressful, the job then becomes a source of strain that can easily lead to a mental injury or illness.

Another factor is the increasing work duties demanded of employees after positions are purposefully not filled or the workforce is actively downsized, resulting in the same amount of work being done by fewer employees. This may not cause employees much harm short-term, but when such changes become long-term or permanent, burnout can certainly rear its ugly head.

While society continues to recognize the stress-related issues that arise with this newer form of work, governments and courts are much slower in acknowledging workplace burnout and stress as real problems that need to be addressed and compensated. As more people learn about burnout and its consequences, it will eventually be recognized in the law as an injury or illness that can drastically impact employee health. Until that day, however, workers will have to keep fighting to have burnout and related illnesses recognized and taken seriously. Only when this happens will legislation and the courts truly protect workers and, therefore, give them the ability to truly advocate for their rights for a safe and healthy workplace.

References

Abu Al Rub, R. (2000) Legal aspects of work related stress in nursing: Exploring the issues. *Official Journal of the American Association of Occupational Health Nurses*, 48(3), 131–5.

Americans with Disabilities Act, 42 U.S.C. § 12101 et seq.

Andolsek v. City of Kirtland, 99 Ohio App. 3d 333, 650 N.E. (2d) 911 (1994).

Armstrong v. Secretary of State for the Home Department, [2001] All E.R. (D.) 83 (Q.B.).

Bailey v. American General, 154 Tex. 430, 279 S.W. 2d 315 (Supreme Ct. 1955).

Barber v. Somerset County Council, [2004] UKHL 13, [2004] 2 All E.R. 385.

Barrett, B. (2004) Notes: Employers' liability for stress at the work place: Neither tort nor breach of contract? *Industrial Law Journal*, 33(4), 343–9.

Bisson, J.I. (2007) Post-traumatic stress disorder. *British Medical Journal*, 334(7597), 789–93.

Blackwell, R. (4 March 2012) Business leaders cite skilled-labour shortage as priority. *The Globe and Mail*. Retrieved from *http://www.theglobeandmail.com/report-on-business/careers/management/c-suite-surveys/business-leaders-cite-skilled-labour-shortage-as-priority/article2358388/*.

Bronstein, A. (2009) *International and Comparative Labour Law: Current Challenges*. Basingstoke, United Kingdom: Palgrave Macmillan/International Labour Organization.

Canada (Royal Canadian Mounted Police) v. Rees, 2005 NLCA 15, 246 Nfld. & P.E.I.R. 79.

Canadian Charter of Rights and Freedoms, Part 1 of the *Constitution Act, 1982*, being Schedule B to the *Canada Act 1982* (UK), 1982, c. 11.

Carter v. General Motors, 361 Mich. 577, 106 N.W. 2d 105 (Supreme Ct. 1960).

Casserley, T.B., and Megginson, D. (2009) *Learning from Burnout: Developing Sustainable Leaders and Avoiding Career Derailment*. Oxford, UK: Butterworth-Heinemann.

Chase, S., Jang, B., and McKenna, B. (16 June 2011) Strike two: Ottawa steps in to end labour dispute at Canada Post. *The Globe and Mail*, A9.

Children's Aid Society of Cape Breton-Victoria v. Nova Scotia (Workers' Compensation Appeals Tribunal), 2005 NSCA 38, 230 N.SR. (2d) 278.

Cox, N. (2004) Workplace stress: What the law says. *Director*, 58(3), 36.

deCarteret, J.C. (1994) Occupational stress claims: Effects on workers' compensation, *Official Journal of the American Association of Occupational Health Nurses*, 42(10), 494–8.

Decision No. 397/92, [1993] O.W.C.A.T.D. No. 70.

Decision No. 1030/89, [1991] O.W.C.A.T.D. No. 903, 20 W.C.A.T.R. 46.

DeFrank, R.S., and Ivancevich, J.M. (1998) Stress on the job: An executive update. *Academy of Management Executive*, 12(3), 55–66.

Dillard, J. (2008) A slaughterhouse nightmare: Psychological harm suffered by slaughterhouse employees and the possibility of redress through legal reform. *Georgetown Journal on Poverty Law and Policy*, 15(2), 391–408.

Directive 89/391/EEC, Health and Safety Framework Directive (1989) Official Journal No. L183/1. Retrieved from: *http://eur-lex.europa.eu/LexUriServ/LexUriServ.do?uri=OJ:L:1989:183:0001:0008:EN:PDF*.

Dobson, M., and Schnall, P.L. (2009) From stress to distress: The impact of work on mental health. In P.L. Schnall, M. Dobson, and E. Rosskam (Eds.), *Unhealthy Work: Causes, Consequences, Cures* (pp. 113–32). Amityville, New York: Baywood Publishing.

Dowling v. Prince Edward Island (Workers' Compensation Board) (1994), 124 Nfld. & P.E.I.R. 358, 7 C.C.E.L. (2d) 157 (P.E.I.C.A.).

England, G. (2008) *Individual Employment Law* (2nd ed.). Toronto, Ontario: Irwin Law.

Entner v. Swissair Transport Co. (24 November 1999), Montreal, Quebec, Doc. No. 500-05-006402-950, J.E. 2000-130.

Federal Employers Liability Act, 45 U.S.C. § 51 et seq.

Greenhouse, S. (15 September 2012) County judge strikes down some restrictions on public unions in Wisconsin law. *The New York Times*, A16. Retrieved from: *http://www.nytimes.com/2012/09/15/us/judge-strikes-parts-of-wisconsin-union-law.html*.

Government Employees Compensation Act, R.S.C. 1985, c. G-5.

Gunderson, M., and Hyatt, D. (Eds.) (2000) *Workers' Compensation: Foundations for Reform*. Toronto, Ontario: University of Toronto Press.

Harthill, S. (2010) The need for a revitalized regulatory scheme to address workplace bullying in the United States: Harnessing the Federal Occupational Safety and Health Act. *University of Cincinnati Law Review*, 78(4), 1250–306.

Hatton v. Sutherland and other Appeals, [2002] EWCA Civ. 76, [2002] 2 All E.R. 1.

Health and Safety at Work etc. Act 1974 (UK), c. 37.

Health and Safety Executive (2007) *Managing the Causes of Work-related Stress* (2nd ed.). Surrey, United Kingdom: Health and Safety Executive. Retrieved from: *http://www.hse.gov.uk/pubns/priced/hsg218.pdf*.

Health and Safety Executive. *Enforcing Health & Safety*. Retrieved from: *http://www.hse.gov.uk/enforce/*.

Health and Safety Executive. *What are the Management Standards?* Retrieved from: *http://www.hse.gov.uk/stress/standards/index.htm*.

Howlett, K. (11 September 2012) Teachers unions vow action after Ontario bill passed. *The Globe and Mail*. Retrieved from: *http://www.theglobeandmail.*

com/news/politics/teachers-unions-vow-action-after-ontario-bill-passed/article4536303/.

Idaho Administrative Code: Occupational Licenses: Rules Governing the Physical Therapy Licensure Board, Rule 24.13.01, subrule 250.06.b.viii (2012) Retrieved from: *http://adminrules.idaho.gov/rules/current/24/1301.pdf.*

Industrial Accident Victims' Group of Ontario and Community Legal Education Ontario. (2006) *Workers' Compensation: A Manual for Workers' Advocates.* (3rd ed.). Toronto, Ontario: Industrial Accident Victims' Group of Ontario and Community Legal Education Ontario. Retrieved from: *http://www.cleo.on.ca/english/wcManual/cover.htm.*

In re Complaint as to the Conduct of Robert J. Loew, 292 Or. 806, 642 P. 2d 1171 (Supreme Ct. 1982).

In re Complaint as to the Conduct of Robert J. Loew, 294 Or. 674, 661 P. 2d 922 (Supreme Ct. 1983).

In re Complaint as to the Conduct of Robert J. Loew, 296 Or. 328, 676 P. 2d 294 (Supreme Ct. 1984).

Ivancevich, J.M., Matteson, M.T., and Richards III, E.P. (1985) Who's liable for stress on the job? *Harvard Business Review*, 63(2), 60–2, 66, 70–2.

Jauregui, M., and Schnall, P.L. (2009) Work, psychosocial stressors, and the bottom line. In P.L. Schnall, M. Dobson, and E. Rosskam (Eds.), *Unhealthy Work: Causes, Consequences, Cures* (pp. 153–67). Amityville, New York: Baywood Publishing.

Labrie et C.S.S.T. – Laurentides et Les Planchers 5 Étoiles Enr. (11 mai 1999), C.S.S.T. No. 113560619-1 (Quebec). Retrieved from: *http://www.jugements.qc.ca/.*

Lippel, K., and Sikka, A. (2010) Access to workers' compensation benefits and other legal protections for work-related mental health problems: A Canadian overview. *Canadian Journal of Public Health*, 101(Supplement), S16–S22.

Meriano, C. (2001) Compensation for stress and mental injury under workers' compensation. *Work*, 17, 257–61.

Maslach, C. (1982) *Burnout: The Cost of Caring.* Englewood Cliffs, New Jersey: Prentice-Hall.

Mealer, M., Burnham, E.L., Goode, C.J., Rothbaum, B., and Moss, M. (2009) The prevalence and impact of post traumatic stress disorder and burnout syndrome in nurses. *Depression and Anxiety*, 26(12), 1118–26.

Mellor, N., Mackay, C., Packham, C., Jones, R., Palferman, D., Webster, S., and Kelley, P. (2011) 'Management standards' and work-related stress in Great Britain: Progress on their implementation. *Safety Science*, 49(7), 1040–6.

Neumann, P., and Sack, J. (2012) *eText on Wrongful Dismissal and Employment Law* (1st ed.). Toronto, Ontario: Lancaster House. Retrieved from: *http://www.canlii.org/en/commentary/wrongfuldismissal/.*

Nova Scotia (Workers' Compensation Board) v. Martin; Nova Scotia (Workers' Compensation Board) v. Laseur, 2003 SCC 54, [2003] 2 S.C.R. 504.

Occupational Safety and Health Act, 29 USCS § 651 et seq.

Ojeda Avilés, A. (2009) The "externalization" of labour law. *International Labour Review*, 148(1–2), 47–68.

Plesner v. British Columbia (Hydro and Power Authority), 2009 BCCA 188, 308 D.L.R. (4th) 624.

Re Morrison Estate, 2003 NSCA 103, 218 N.S.R. (2d) 53.
Reidy v. Travelers Ins. Co., 928 F. Supp. 98 (D. Mass. 1996).
Reidy v. Travelers Ins. Co., 107 F. 3d 1 (Table) (1st Cir. 1997).
Sandborn, T. (26 July 2012) Despite new rules, mentally injured and bullied workers still disadvantaged: Critics. *The Tyee*. Retrieved from: *http://thetyee.ca/News/2012/07/26/Mentally-Injured-Workers/*.
Shain, M. (2009) *Stress at Work, Mental Injury and the Law in Canada: A Discussion Paper for the Mental Health Commission of Canada*. Ontario and Alberta: Mental Health Commission of Canada. Retrieved from: *http://www.mentalhealthcommission.ca/English/system/files/private/Workforce_Stress_at_Work_Mental_Injury_and_the_Law_in_Canada_ENG_0.pdf*.
Shain, M. (2010) *Tracking the Perfect Legal Storm: Converging Systems Create Mounting Pressure to Create the Psychologically Safe Workplace*. Ontario and Alberta: Mental Health Commission of Canada. Retrieved from: *http://www.mentalhealthcommission.ca/English/system/files/private/Workforce_Tracking_the_Perfect_Legal_Storm_ENG_0.pdf*.
Shain, M., Arnold, I., and GermAnn, K. (2011) *The Road to Psychological Safety: Legal, Scientific and Social Foundations for a National Standard for Psychological Safety in the Workplace*. Ontario and Alberta: Mental Health Commission of Canada. Retrieved from: *http://www.mentalhealthcommission.ca/English/system/files/private/Workforce_The_Road_to_Psychological_Safety_ENG_0.pdf*.
Shortt, S.E.D. (1995) The compensability of chronic stress: A policy dilemma for the Ontario workers' compensation board. *Canadian Public Policy*, 21(2), 219–32.
Smith, C.M. (1995) Workers' compensation: Compensation for mental-mental claims. *American Journal of Trial Advocacy*, 19(1), 229–32.
Slaight Communications Inc. v. Davidson, [1989] 1 S.C.R. 1038, 93 N.R. 183.
Stafford and Rural Homes Ltd. v. Hughes (2009), 107 B.M.L.R. 155 (U.K.).
State ex rel. Oklahoma Bar Ass'n v. Schraeder, 51 P. 3d 570 (Okla. Supreme Ct. 2002).
Stefan, S. (1998) "You'd have to be crazy to work here": Worker stress, the abusive workplace, and title 1 of the ADA. *Loyola of Los Angeles Law Review*, 31(3), 795–846.
Svirsky, G. (1998) The division of labour: An examination of certification requirements. *Osgoode Hall Law Journal*, 36(3), 567–96.
The Health and Safety (Consultation with Employees) Regulations 1996, S.I. No. 1996/1513.
The Management of Health and Safety at Work Regulations 1992, S.I. No. 1992/2051.
The Management of Health and Safety at Work Regulations 1999, S.I. No. 1999/3242.
The Workplace (Health, Safety and Welfare) Regulations 1992, S.I. No. 1992/3004.
Tucker, A. (2010) A matter of fairness: How denying mental-mental claims frustrates the central purposes of workers' compensation law. *The Journal of Legal Medicine*, 31(4), 467–84.

Van Alphen, T. (26 January 2012). 'Moment of truth' for unions; Secret paper on massive proposed merger says labour groups must reinvent themselves or die. *The Toronto Star*, A1.

Walker v. Northumberland County Council (1994), [1995] 1 All ER 737 (Q.B.).

Wallace v. United Grain Growers Ltd., [1997] 3 S.C.R. 701, 123 Man. R. (2d) 1.

West Virginia Legislative Rule: Bureau for Public Health: Alzheimer's/Dementia Special Care Units and Programs, Rule 64–85, subrule 4.1.c.10 (2006). Retrieved from: *http://apps.sos.wv.gov/adlaw/csr/readfile.aspx?DocId=8999andFormat=PDF.*

Workers' Compensation Act, R.S.B.C. 1996, c. 492.

Workers' Compensation Act, R.S.P.E.I. 1988, c. W-7.1.

Workers' Compensation Act, S.N.S. 1994–1995, c. 10.

Workers' Compensation Board of Nova Scotia. Compensability of Stress as an Injury Arising out of and in the Course of Employment – Government Employees Compensation Act (GECA): Policy 1.3.6. (2005). Retrieved from: *http://www.wcb.ns.ca/policy/index_e.aspx?DetailID=1511.*

Workplace Health, Safety and Compensation Act, R.S.N.L. 1990, c. W-11.

Workplace Health, Safety & Compensation Commission of Newfoundland and Labrador. Policies & Procedures: Entitlement, Mental Stress, Policy No. EN-18 (1999) Retrieved from: *http://www.whscc.nf.ca/policiesandprocedures.whscc.*

Workplace Safety and Insurance Act, 1997, S.O. 1997, c. 16, Sch. A.

WorkSafeBC (1998) *Compensation for Chronic Stress: A Briefing Paper.* Retrieved from: *http://www.worksafebc.com/regulation_and_policy/archived_information/royal_commission_briefing_papers/assets/pdf/stress.pdf.*

WorkSafeBC (2009a) *Discussion Paper: Chronic Stress.* Retrieved from: *http://www.worksafebc.com/regulation_and_policy/policy_practice_consultative_committee/meeting_calendar/august09/assets/chronicstress.pdf.*

WorkSafeBC (2009b) Rehabilitation Services & Claims Manual, Volume II, Policy 13.30. Retrieved from: *http://www.worksafebc.com/publications/policy_manuals/rehabilitation_services_and_claims_manual/volume_ii/assets/pdf/rscm_ii_sum0907.pdf.*

WorkSafeNB: Federal Government Employees, Policy 21-011, subrule 3.1. (2011) Retrieved from: *http://www.worksafenb.ca/pdf/resources/policies/21-011.pdf.*

Zorn-Smith v. Bank of Montreal, [2003] O.T.C. 1060, 31 C.C.E.L. (3d) 267 (Superior Ct.).

7

Gender, burnout and work-related stress

Abstract: This chapter discusses gender, burnout and work-related stress. Specifically it looks at the role gender plays with regard to work–life balance, pay, career advancement, stress, mental health and physical health.

Key words: burnout, gender, work-related stress, work–life balance, pay, career advancement, mental health, physical health.

Introduction

When we turn to the matter of gender and burnout, we encounter two specific problems. The first is that (as described in Chapter 2) researchers have turned up mixed results as to whether women suffer from burnout more readily or more often than men. In some studies, women appear to be more prone to burnout; in others it appears gender makes no difference.

The second problem, as discussed in Chapter 1, is that burnout is not a recognized mental health problem in countries which rely on the *Diagnostic and Statistical Manual of Mental Disorders,* Fourth Edition (DSM-IV). In these countries, burnout is likely to have been diagnosed as stress or depression. As a result, it is important to survey the literature on workplace stress and depression in these parts of the world since individuals suffering from burnout are likely to surface in these results but not to have been identified specifically.

Given this problem of identification and diagnosis, it is little wonder that the research is not clear on whether men or women are more burned out. It is also not clear on whether women or men typically experience more work-related stress. Some researchers, for example, claim that men are more acutely affected by work-related stress. At the 2008 British Psychological Society's occupational psychology conference in Stratford-upon-Avon, business psychologist Lucy Watt explained that, based on her studies, men suffer from work-related stress more than women, but, remarkably, experience fewer health-related problems as a result. She explained that men seemed better able to buffer themselves from the harmful effects of stress (Jones, 2008).

In contrast, research presented in 2009 by Dr Brendan Burchell from the University of Cambridge's Department of Sociology indicated that while men are indeed more stressed, depending on the circumstances, they are less able than women to buffer themselves from the effects of stress. Burchell's studies revealed that men who feel insecure about their jobs are more likely than women to suffer from depression and anxiety. Burchell analyzed data from a survey of 300 UK employees along with the results from thousands of others through the British Household Panel Survey, funded by the Economic and Social Research Council (ESRC). He found that while men claim to be less worried about losing their jobs, they are, in fact, more stressed, worried and depressed than women (University of Cambridge, 2009).

For the most part, though, results like these showing that men suffer more work-related stress than women are not as common as studies which indicate that women are actually more stressed. A recent poll taken of 2,000 Britons by the Stroke Association, for example, found that twice as many women as men were experiencing high levels of stress. Approximately one-in-five women and one-in-ten men felt that their stress levels were out of control. The sources of stress were jobs, job insecurity and financial problems (Jones, 2011).

Along the same lines, a 2010 survey of 29,338 men and women in India, China, Brazil, Russia, America, Britain, Australia, Canada, Denmark, Finland, France, Germany, Italy, Japan, Mexico, the Netherlands, Spain, Sweden and Switzerland conducted by Kenexa Research Institute, found that women experienced more workplace stress than men. Stress levels were generally between 6 to 10 percent higher for women than they were for men depending on the type of job. Work–life balance, fair pay, supportive management, having a career path, and the type of work were all issues reported by the survey participants which directly contributed to their levels of stress ("Women suffer," 2010).

This chapter looks at the research on gender and workplace stress. Several surprising and some alarming results have emerged over the last several decades which indicate that women not only experience workplace stress more acutely, but also suffer serious health problems, and sometimes even put their lives in danger because of their jobs.

Gender, burnout and work-related stress

Gender and work–life balance

Why are women more stressed? The explanation is that women, more than ever, are not only struggling with the stresses of the job but also with increasing demands at home including caring for elderly parents and raising children. As Paul Farmer, the chief executive of Mind, a mental health charity, said: "Undoubtedly, there is a combination of factors affecting women which mean they face greater challenges, maintaining work–life balance and caring responsibilities. Where there is increasing pressure at work and at home, it is often women who are expected to bear the brunt" (Laurance, 2009).

In single-parent households, the situation is particularly bleak. One study of single- and dual-parent households found the single mothers suffered from "role overload," that is, they worked constantly either as a parent or a wage-earner with little or no downtime. As a result, they were "three times more likely to experience severe depression than the married women in the study. They were 80 per cent more likely to have chronic health problems and 40 per cent more likely to suffer from flu or colds" (Mironowicz, 1993). Frequent illness indicated a deep level of physical and mental exhaustion on the part of single mothers who were described as "worn out," and they were 50 percent more likely than married mothers to use health services to deal with both depression and physical health problems.

This does not mean, however, that women who are married are stress free. Time pressure is a big problem for couples, and significant changes have occurred in the homes of married couples since the 1970s which have added to the stress. In 1970, for example, among two-thirds of married couples, one of those individuals – often the wife – stayed at home. The stay-at-home spouse had a full day to see to domestic responsibilities while the other spouse earned a salary. Today, however, that number has fallen to 40 percent which means that more spouses must tackle these domestic obligations after a full day at work (Parker-Pope, 2010). In addition, the average work week is longer for both spouses, having risen from 52.5 hours a week in 1970 to 63 hours in 2009 (Parker-Pope, 2010).

Women take on the lion's share of domestic responsibilities. According to one US survey, women spend 28 hours a week on housework, while men claim to spend approximately 16 hours. There is also a vast difference in perception between the sexes as to how much time men spend on childcare. Close to half (49 percent) of men surveyed said they spent an equal amount of time or more on childcare than their spouse. In contrast, the women surveyed believed that only 31 percent of those men spent the amount of time they reported. More than half of all men also claimed to spend an equal amount of time or more on housework while 70 percent of their wives disagreed (Parker-Pope, 2010).

Even though women and men may not agree on the percentages, and even though men may not be shouldering half of the domestic responsibilities, it is clear that more men than ever before are troubled by the competing demands of work and family. Studies indicate that the number of men who experience this work–family conflict has risen noticeably since the 1970s. In 1977, for example, a US Department of Labor survey revealed that 34 percent of men experienced "some" or "a lot" of work–family conflict. In 2008, New York-based Work and Family Institute updated that study and found that 49 percent of working men with families were now reporting the same thing (Healy, 2011). The reasons for this include modern expectations that men will be more involved with their families. At the same time, wages are stagnating and hours of work, along with worries about job insecurity, have risen significantly.

So, while women in two-parent households still shoulder the burden of housework and child raising, men are not entirely stress free. Societal pressure for men to be more involved in family life than in previous decades means that work–home time pressures are increasingly becoming more of a problem for men as well as women.

Gender, equal pay and barriers to career advancement

Women are greater participants in the workforce than they have ever been in history and the traditional view of the husband as the primary breadwinner while the wife earns "pocket money" has been completely turned on its head. In the United States, for example, women hold half the jobs in the economy and were the "main or co-breadwinners in two-thirds of American families" (Cobble, 2010). In contrast, men represented "a majority of the jobless" (Cobble, 2010). Because men were hardest hit by the recession in the US, now 38 percent of women either make more money than their husbands or are the sole breadwinners in the household (Eichler, 2012).

The notion that women are less engaged in the workforce once they have children is also changing. In Canada, 70 percent of women with children who are still at home have jobs (Eichler, 2012).

Nonetheless, the expansion of women's participation in the workforce has not led to corresponding significant improvements in the conditions of work. Instead, men still outpace women in terms of wages even for equal work. In the US, the gender pay gap is 23 percent (Cobble, 2010). In Canada, men earn on average 20 percent more than women ("Canada gender pay gap," 2010) while in the UK that figure is 19.5 percent (Jones, 2011).

Low pay has been and remains a women's issue. A 2012 work-stress survey conducted by Harris Interactive on behalf of Everest College found that 73 percent of Americans felt stressed in the workplace, and low pay was the primary reason (11 percent). Close to double the amount of women (14 percent) ranked low pay as their top stressor compared with men (8 percent).

In addition, the female participants (11 percent) said that they were stressed because the job they were in was not their ideal career compared with men (5 percent) (Harris Interactive, 2012). Some of this has to do with the greater barriers women encounter in the job market, not only in securing certain jobs, but also in moving up the job ladder.

Gender, mental health and work-related stress

Both men and women are more likely to suffer from psychiatric disorders as a result of high pressure (long hours, time pressures) on the job. One study published in 2007 found that 45 percent of young adults in their early 30s with no pre-employment history of depression or anxiety could attribute the onset of psychiatric disorder to the excessive demands of their job. In other words, job stress was responsible for new psychiatric disorders in both men and women who had been mentally healthy up to that point (Melchior et al., 2007). Women in high pressure jobs were 75 percent more likely that women in low stress jobs and men were 80 percent more likely than men in low stress jobs to suffer from depression or anxiety.

Depression was also the focus of a recent British study which found that employees who spent excessive amounts of time at work were more likely to suffer from depression than those employees who worked a typical day. Individuals who worked 11-plus hour days (or 55-plus hours a week) were two and a half times more likely to report depression than those who worked a typical seven to eight hour day. In reaching this finding, researchers took into account the effect of the various socio-demographic factors of these employees and also the extent to which they lived a healthy or unhealthy lifestyle (Virtanen et al., 2012).

Women in these studies are generally more likely to suffer from depression than men. Researchers surveyed over 2000 British civil servants from 1991 to 1993 and followed up again six years later from 1997 to 1999. They found that women, younger workers, and low ranking employees with moderate alcohol use were most at risk for the onset of depression. Higher ranking men in the workplace did not seem particularly prone to depression; in contrast, higher ranking women were. Researchers surmised that the difference between men and women at this level of management had to do with the prevalence or absence of ongoing responsibilities at home ("A long work day," 2012).

Work-related depression was also found to affect more women than men in a 2010 University of Melbourne and Victoria Health report. Researchers noted that 17 percent of women (compared with 13 percent of men) suffered from work-related depression as a result of stress on the job.

Finally, an extensive study of 30 European countries found that women between the ages of 25 and 40 are three to four times more likely than men to be depressed. Women in all age groups are two and a half times more likely to be more depressed than men. The reason, according to researchers, has to do with the stresses of balancing work and home commitments. Women not only suffer from depression earlier in life than they did in the 1970s but the rate has doubled since that time (Collins, 2011).

Hans Ulrich Wittchen, a professor at Dresden University of Technology in Germany, and one of the authors of the study, noted that

> [m]arriage appears to reduce the risk of depression in males, for females it increases the risk. In females, you see these incredibly high

> rates of depressive episodes at times when they sometimes have their babies, where they raise children, where they have to cope with the double responsibility of job and family. (Collins, 2011)

The study also found that rates of depression in men had also risen since the 1970s (although not to the same extent) which Professor Wittchen attributed to an insecure job market, worries about job loss and higher divorce rates (Smyth, 2011).

Finally, research on how men cope with their job-related stress and depression and whether they seek help has been somewhat more alarming. One-third of all British males, for example, report they use alcohol to stop themselves from thinking about work (Tozer, 2006). Twenty-two percent of the males in this particular survey reported suffering from work-related depression while 35 percent said that work was responsible for problems sleeping. One in ten respondents saw work as responsible for a decreased interest in their family or significant other and 15 percent felt that work had disrupted their sex lives. Despite the impact on their family and health, however, men continue to be more reluctant than women to seek help (Tozer, 2006).

This is borne out by statistics in the UK which indicate that women are seeking help for stress in greater numbers than ever before and outpacing men. In 2009, for example, women made up 56 percent of all the referrals to National Health Service specialist psychiatric services for outpatient and inpatient treatment (Laurance, 2009).

Gender, physical health and work-related stress

The American Psychological Association has also reported that women are more likely to be stressed than men. Work was cited as being the second most significant source of stress (money was number one) and, once again, women reported higher average levels of stress than men. Of the women surveyed, 88 percent felt that stress was having an adverse impact on their health compared with 78 percent of men (American Psychological Association, 2012).

Work-related stress is not just a problem for women, it is also a serious problem for men. One study of job stress among young Finns (mean age 32) found that even though young women tended to report more work-related stress (35.6 percent versus 27.4 percent), it was young men who showed signs of early atherosclerosis (a build up of plaque inside the arteries) which placed them at risk for heart attacks and strokes. The young men were 29 percent more likely than women to have increased thickness in their arteries. This was true even when other unhealthy lifestyle factors (drinking, smoking, obesity, etc.) were taken into account (Hintsanen et al., 2005).

Men were also found to be at risk for certain health problems in a Danish study which in 2001 followed up with a group of healthy men first identified in 1970. Researchers found that high stress at work was a significant predictor of stroke, and that 10 percent of all strokes could be attributed to job stress, while 90 percent were attributable to other factors such as smoking and high blood pressure. This was true, however, only for men in higher social classes – i.e., those with higher education and job status. Little or no association between stroke and high job stress was found in men in the lowest social classes – i.e., those individuals with little education and low ranking jobs (Suadicani et al., 2011).

These studies tell us that job stress is clearly unhealthy for men. The same is true for women. Stress has consistently been linked with poorer health in women, and, for the most part its effects seem even more severe than they are for men. One study done in Sweden of male and female clerical workers and middle-managers, for example, found that both the male and female managers listed a "heavy workload, time pressure and responsibility for others" (Frankenhaeuser et al., 1989) as the main reasons for stress on the job. Women, however, felt additional sources of stress as compared to the men: one-third of the women felt that they had to perform better than the men merely to be considered equals. Also, while male managers showed the greatest job satisfaction and reported high levels of personal control, their counterparts – female managers – felt they had less control and less support from the managers above them.

The most significant difference between the men and women managers in this Swedish study was that blood pressure remained high for the women after work, while the blood pressure in the men decreased after 5 p.m. According to researchers, this indicates that women continue to experience stress after work as a result of the demands made of them at home. The study indicated that high demands at work, followed by more demands in their domestic life, left this group unable to unwind in the same way as their male counterparts. It also put the women at a greater risk for cardiovascular disease in the future.

Ten years after this study, in 1999, researchers at Laval University published a paper which showed similar results in their investigation of working women. Here, the link between high blood pressure in university- and non-university-educated women who were raising children was examined. University-educated women working in high-demand jobs had significantly higher blood pressure levels than their non-university-educated counterparts with similar domestic responsibilities. The blood pressure of the university-educated women remained high not only after the workday had ended, but throughout the evening and into the night, putting this group at risk for heart attacks and strokes (Brisson et al., 1999).

Shortly after the Laval Study was published, researchers from the Harvard School of Public Health published research which analyzed data collected from 21,290 nurses. The participant nurses had filled out surveys in both 1992 and 1996, and researchers looked at information about their jobs, along with changes in health (particularly any decline in health) over the four years between those two questionnaires.

They discovered that women in jobs with high demands (e.g., a relentless pace, not enough time to perform the work, too much work) but low ability to participate in decision making or to have some say in how the job was to be done had experienced the greatest decline in their physical and mental health. Low social support also contributed to poor mental and physical health (Cheng et al., 2000). "Some jobs are inherently stressful," said Ichiro Kawachi, Associate Professor of Health and Social Behavior and one of the study's authors. "What we found is that regardless of how healthy or resilient a person is, the job stress can affect her health" ("Major study," 2000).

The researchers also noted that while previous research had linked job stress to heart disease and various other health problems, the current study revealed that health problems continue to get worse over time if job stress is unabated ("Major study," 2000).

In addition to heart disease and risk of stroke, a study published in 2004 found a link between job stress and health matters which are specifically female such as premenstrual symptoms. Researchers surveyed over 6,000 active-duty women working across all branches of the US military. Sixty-nine percent of the women had experienced symptoms in the three months prior to the study, and job stress was found to be the strongest predictor of these symptoms (Hourani et al., 2004). Researchers explained that more than 40 percent of women in the military report experiencing "a great deal" or "a fairly large amount" of stress as part of their jobs.

Pregnancy and health of the child has also been found to be more problematic for women in jobs with long hours. Dutch researchers who focused on 7,000 pregnant women in Amsterdam found that women who worked in high stress jobs for more than 32 hours a week were at greater risk for a pregnancy condition called pre-eclampsia. Pre-eclampsia involves high blood pressure and damage to the maternal endothelium (the cells lining the blood and lymphatic vessels), liver and kidneys. The women reporting high stress on the job also gave birth to babies who were low birth-weight (a similar birth-weight, in fact, to that of women who smoked) and who cried excessively ("Women in high stress jobs," 2006).

Like the Harvard School of Public Health study discussed above, a number of these studies have followed a population group over the course of several years to determine how stress is affecting their health. A Danish study which began in 1993, for example, collected data from 12,116 employed female nurses and followed this group for 15 years. The study found workload reported as being much too high could predict ischaemic heart disease (IHD) at a later date (Allesøe et al., 2010). Surprisingly, the results were even more pronounced among younger nurses.

A long term study in Britain (which followed 10,000 civil servants for 14 years) found that employees in high stress jobs had a better chance than those in low stress positions to be obese, and have higher cholesterol and blood pressure (known as metabolic syndrome) – all factors which put workers at risk for heart disease and diabetes (Chandola et al., 2006). While stressed men were twice as likely as their non-stressed male counterparts to develop the syndrome, women were five times more likely to have the syndrome when compared to men. According to experts, "one possible explanation is that prolonged exposure to work stress may affect the nervous system. Alternatively, chronic stress may reduce biological resilience and thus disturb the body's physiological balance (homoeostasis)" (Boseley, 2006).

The results of another long term study were presented in 2010. Dr. Michelle Albert of Brigham and Women's Hospital in Boston, speaking at the American Heart Association in Chicago, explained that her team of researchers had also discovered a link between job strain and cardiovascular disease. The study, sponsored by the US National Heart, Lung and Blood Institute, analyzed data gathered from over 17,000 women working in the health profession over a ten year period.

What the study found was that women who endured high demands but low control in their job had a much greater risk of heart attack (88 percent higher in women with high job stress), stroke, bypass surgery (43 percent higher) and death (Ubelacker, 2010). In total, women's risk of suffering from some form of cardiovascular disease was 40 percent higher as a result of job stress.

Researchers in this same study noted that the fear of losing one's job increased the factors which could lead to heart disease including high blood pressure and cholesterol and being overweight. Stress was also responsible for the release in the body of stress hormones such as cortisol which does damage by narrowing blood vessels, and affecting heart rate and blood pressure (Ubelacker, 2010).

What many of the foregoing studies have in common is that they have based a number of their assumptions on Karasek's "job strain" model. This model posits that jobs with high demands and low autonomy or decision-making power result in the greatest health risks from stress. The US National Heart, Lung and Blood Institute study above, for example, found a link between high demand–low control jobs and poor health outcomes. One study in which data was compiled from 22,000 women over the course of ten years, however, indicated that this is not always the case.

In this study, researchers at Harvard and Yale found, much like the studies above, that women with stressful jobs were at much greater risk for serious physical health problems than their less stressed female counterparts. Results indicated that women suffering from job stress were 67 percent more likely than women with less stressful jobs to have a heart attack and 38 percent more likely to have a cardiovascular event such as a stroke (Slopen et al., 2012). Where this study differed from a number of the others which had come before it was that researchers discovered that high demand jobs, whether or not women had control over their work, resulted in greater health problems. In other words, high demands in the workplace were enough to lead to health problems. In contrast, men with jobs in which they had high demands and high control were found not to experience the same cardiovascular problems as women (Pearson, 2012).

Conclusion

Research is clear that job stress is mentally and physically unhealthy for men and women. Members of both sexes are more likely to develop psychiatric disorders when exposed to ongoing and high levels of stress in the workplace, and both sexes find themselves at greater risk for heart problems and strokes. Women, however, have been shown in most studies to suffer these effects more acutely than men particularly as they move up the ladder in an organization. As research suggests, women are more likely to suffer serious health problems when exposed to the stresses of the job, most likely because they are unable to unwind at home once their work day has come to a close.

References

Allesøe, K., Hundrup, Y.A., Thomsen, J.F., and Osler, M. (2010) Psychosocial work environment and risk of ischaemic heart disease in women: The Danish nurse cohort study. *Journal of Occupational and Environmental Medicine*, 67(5), 318–22.

American Psychological Association (11 January 2012) *Stress in America: Our health at risk*. Retrieved from *http://www.apa.org/news/press/releases/stress/2011/final-2011.pdf*.

Boseley, S. (20 January 2006) Heart disease and diabetes linked to stress in the office: Researchers study 10,000 civil servants for 14 years: Women five times more likely to have risk factors. *The Guardian*, 11.

Brisson, C., Laflamme, N., Moisan, J., Milot, A., Mâsse, B., and Vézina, M. (1999) Effect of family responsibilities and job strain on ambulatory blood pressure among white-collar women. *Psychosomatic Medicine*, 61(2), 205–13.

Canada Gender Pay Gap among worst in OECD (8 March 2010) *Toronto Sun*. Retrieved from *http://www.torontosun.com/money/2010/03/08/13155136.html*.

Chandola, T., Brunner, E., and Marmot, M. (2006) Chronic stress at work and the metabolic syndrome: Prospective study. *British Medical Journal*, 332(7540), 521–5.

Collins, N. (5 September 2011) Women more than twice as likely to be depressed. *The Telegraph*. Retrieved from *http://www.telegraph.co.uk/health/8740278/Women-more-than-twice-as-likely-to-be-depressed.html*.

Cheng, Y., Kawachi, I., Coakley, E.H., Schwartz, J., and Colditz, G. (2000) Association between psychosocial work characteristics and health functioning in American women: Prospective study. *British Medical Journal*, 320(27), 1432–6.

Cobble, D.S. (3 January 2010) A not so new deal. *Tulsa World*, G3.

Eichler, L. (18 August 2012) Why we should still mind the wage gap. *The Globe and Mail*, B14.

Frankenhaeuser, M., Lundberg, U., Fredrikson, M., Melin, B., Tuomisto, M., Myrsten, A.L., Hedman, M., Bergman-Losman, B., and Wallin, L. (1989) Stress on and off the job as related to sex and occupational status in white-collar workers. *Journal of Organizational Behavior*, 10(4), 321–46.

Harris Interactive (15 August 2012) Three-quarters of Americans stressed by something at work; Everest College's annual work stress survey finds low pay at top of list for second consecutive year; Fear of losing job drops considerably. *GlobeNewswire*.

Healy, M. (10 July 2011) Competing work/family demands stress men too, study finds. *Charleston Gazette*, P4F.

Hintsanen, M., Kivimäki, M., Elovainio, M., Pulkki-Råback, L., Keskivaara, P., Juonala, M., Raitakari, O.T., and Keltikangas-Järvinen, L. (2005) Job strain and early atherosclerosis: the cardiovascular risk in young Finns study. *Psychosomatic Medicine*, 67(5), 740–7.

Hourani, L.L., Yuan, H., and Bray, R.M. (2004) Psychosocial and lifestyle correlates of premenstrual symptoms among military women. *Journal of Women's Health*, 13(7), 812–21.

Huge cost of stress (10 October 2010) *Sunday Telegraph*, 40.
Jones, A. (11 January 2008) Men suffer more work-related stress. *Press Association National Newswire*.
Jones, A. (23 November 2011) Gender pay gap falls. *The Independent*. Retrieved from *http://www.independent.co.uk/news/uk/home-news/gender-pay-gap-falls-6266503.html*.
Jones, C. (5 January 2011) Many start 2011 already stressed by job and financial worries. *The Western Mail*, 7.
Laurance, J. (25 March 2009) More women suffering from stress problems. *The Independent*, 2–3.
A long workday is a short cut to depression: Research (5 March 2012) *Daily News*, 5.
Major study of women at work links job stress with poor, failing health (2000) *New Therapist*, 7.
Melchior, M., Caspi, A., Milne, B.J., Danese, A., Poulton, R., and Moffitt, T.E. (2007) Work stress precipitates depression and anxiety in young, working women and men. *Psychological Medicine*, 37(8), 1119–29.
Mironowicz, M. (6 July 1993) Single mothers under stress from work and child care. *The Windsor Star*, C3.
Parker-Pope, T. (21 June 2010) Equal at last: Men also feel stress over work vs. family. *International Herald Tribune*, 4.
Pearson, C. (20 July 2012) Job stress could raise women's heart attack risk by 70 percent. *Huffington Post*. Retrieved from *http://www.huffingtonpost.com/2012/07/20/job-stress-women_n_1687276.html*.
Picard, A. (29 March 1999) Balancing work and home tasks raising women's blood pressure: Study finds high-stress jobs and child-rearing responsibilities are increasing chances of having stroke and heart disease. *The Globe and Mail*, Metro, A6.
Slopen, N., Glynn, R.J., Buring, J.E., Lewis, T.T., Williams, D.R., and Albert, M.A. (2012) Job strain, job insecurity, and incident cardiovascular disease in the Women's Health Study: Results from a 10-year prospective study. *PLOS ONE* 7(7). Retrieved from *http://www.plosone.org/article/info%3Adoi%2F10.1371%2Fjournal.pone.0040512*.
Smyth, C. (5 September 2011) Depression rate among women soars, research finds. *The Times*.
Suadicani, P., Andersen, L., Holtermann, A., Mortensen, O.S., and Gyntelberg, F. (2011) Perceived psychological pressure at work, social class, and risk of stroke. *Journal of Occupational and Environmental Medicine*, 53(12), 1388–95.
Tozer, J. (8 June 2006) Men who drink to forget the stress of work. *Daily Mail*, 21.
Ubelacker, S. (15 November 2010) Job strain jeopardizes women's health; Stress boosts heart-attack risk by 88%, study finds. *The Hamilton Spectator*, A8.
University of Cambridge (2009) Men hit harder by credit crunch stress. Retrieved from *http://www.cam.ac.uk/research/news/men-hit-harder-by-credit-crunch-stress/*.
Virtanen, M., Stansfeld, S.A., Fuhrer, R., Ferrie, J., and Kivimäki, M. (2012) Overtime work as a predictor of major depressive episode: A 5-year follow-up

of the Whitehall II Study. *PLOS ONE*, 7(1). Retrieved from *http://www.plosone.org/article/info:doi/10.1371/journal.pone.0030719.*

Women in high stress jobs should cut working hours during pregnancy: Research (21 April 2006) *Agence France Presse.*

Women suffer more stress at work (22 July 2010) *Times of India*. Retrieved from *http://articles.timesofindia.indiatimes.com/2010-07-22/bangalore/28287797_1_work-stress-fewer-women-women-fight.*

8

Symptom recognition and preventing burnout

Abstract: This chapter discusses how workers and their employers can recognize and prevent burnout. The chapter uses Leiter and Maslach's six job domains – workload, control, reward, community, fairness, and values – to frame the discussion on how to decrease sources of stress in each of these areas. Additional information is offered on how to improve mental and physical resilience.

Key words: burnout, symptom recognition, prevention, workload, control and reward, community, fairness and values, time management, resilience, lifestyle and vacations.

Introduction

Maslach et al. (2001) note that burnout, unlike depression or stress, is generally job-related and situation specific. What this means is that something is happening at work which is changing the way an individual thinks, feels and reacts within the work environment. Nonetheless, recognizing early warning signs of burnout is not easy, particularly as those early indications can be "confused, sporadic and short-lived" (Glicken and Janka, 1982, p. 69).

Given the confused and sporadic warning signs, it is no wonder that employees at risk generally ignore, do not notice, or do not understand what is going on (Casserley and Megginson, 2009). As Harold Sala (2009) describes it,

> [s]uppose you are driving down the highway in your car and the red light on your instrument panel indicates that your oil pressure is

> down. The engine still sounds good. When you gun the throttle, it responds normally. What do you do? You can either continue driving and ignore the light, or you can pull over, investigate, and determine what is wrong. Even if you choose to ignore the warning light, your automobile will continue to perform for a while just as it always has. But eventually, you'll find yourself at the side of the road with smoke pouring out from under the hood of the car. You are finished – at least for the moment. (p. 218)

Yet burnout's red light on the instrument panel has been well documented in the literature. Employees who are burning out are likely, at first, to be overachievers and, later, to show inconsistent or decreased productivity. They become lethargic, withdraw from others, show a lack of direction, and indicate a desire to change jobs even when there is no opportunity for greater pay or status. They are also more likely to be preoccupied with issues that seem beyond the scope of their immediate employment including questions involving the meaning of life or, somewhat more mundanely, whether a new job or a change in their personal living arrangements is likely to result in greater happiness (Glicken and Janka, 1982).

Another sign of burnout is that employees will often respond to various work-related situations inappropriately, and some of their reactions are likely to inhibit the possibility of their own personal success in the future (Glicken and Janka, 1982, pp. 69–70). In other words, the way employees act or respond to their own predicament puts them at risk of derailing their career at some point in the future.

If employees have difficulty recognizing the early signs of burnout, they also will have trouble avoiding it. Yet early detection or recognition is vital if employees are to take steps to cope with the various sources of burnout.

Coping has been defined as "*efforts* to master conditions of harm, threat, or challenge when an automatic response is not readily available. Coping in itself does not imply *success* but *effort*. It is the link between stress and adaptation" (Pines and Aronson, 1988, p. 144; italic in original).

Leiter and Maslach (2005) note that the literature on coping techniques and burnout prevention tends to focus too heavily on individual-centred solutions rather than on changing the workplace. Sadly, the truth seems to be that workplaces are difficult to change, so employees usually must rely on personal coping skills.

Given this fact, burnout studies over time have sought to find out which methods individuals employ (e.g., setting boundaries, practicing time management, relaxation techniques, engaging in physical exercise, etc. See Gupta et al., 2012) to combat burnout. They have also explored just how effective the individuals in question feel those methods to be.

Some studies, for example, have participants try certain coping methods and strategies and then look at pre- and post-study burnout scores to determine if scores are lowered after the study. Research has been done on the use of cognitive behavioral training, counseling (van Dierendonck et al., 1998), adaptive coping (Rowe, 2000), psychotherapy (Salmela-Aro et al., 2004), relaxation training (Van Rhenen et al., 2005), mindfulness (Moon, 2009), laughter therapy (Kanji et al., 2006) and so on. While many of these strategies have been found to be effective in reducing burnout scores, keeping those scores down has proven more difficult after six months to a year (Awa et al., 2010). In other words, at first, scores recorded after participants have used a successful coping method may indicate a reduced rate of burnout, and even stay that way for some time. Inevitably though, if changes do not occur in the workplace, those rates of burnout begin to rise once again.

Why is this the case? Research has shown that burnout is generally a consequence of a dysfunctional relationship between two parties – the individual and the workplace. Techniques for burnout prevention, therefore, will only be partially successful if the individual has little or no control over the sources of stress in the workplace. When individuals alone seek to change the troubling elements within their control, but the organizational environment remains static, problems will most likely remain. Clearly then, the optimum way to combat burnout symptoms is to have both the organization and the individual participate in burnout interventions (Schaufeli and Enzmann, 1998; Maslach et al., 2001).

Realistically speaking, however, we know that many organizations are monolithic and that matters such as the amount of work, the work schedule, the supervisor and the very nature of the work itself are not negotiable. Because of this, such workplaces often cannot, or will not, give workers the autonomy necessary to make key changes in their jobs.

Nonetheless, the picture is not entirely bleak. There are steps, small and big, which individuals can take, not only to take care of themselves personally but also to attempt to change their work environment. There are also factors that more enlightened organizations can consider to make sure their employees are not at risk. Leiter and Maslach (2005) focus on six job domains – workload, control, reward, community, fairness, and values – and discuss how to work out a specific action plan in order to repair the problems arising in each of these domains. Their suggestions, along with the studies and suggestions of other experts in this area, are discussed, first from the standpoint of the individual and then from the standpoint of the organization, under these six job domain headings.

Leiter and Maslach (2005), along with many other experts in this area, recommend building resilience by focusing on health. As such, we also provide in this chapter recommendations for healthier living such as exercising and eating properly, which are key to keeping the body and mind healthy in situations involving high stress.

What the employee can do: burnout prevention in six job domains

Job domain 1: workload

Of the six job domains in which problems arise, workload is number one. Leiter and Maslach (2005) along with other authors note that self-awareness and self-monitoring are critical in preventing burnout, and when tackling the problem of workload, workers need to know what fires them up about the job and what, on the other hand, drains them of energy. Is it simply a case of too much work? Are deadlines too severe? Is it the type of work? Is the employee expected to be available at all times? Coming up with a solution is easier once the right questions have been asked and the specific problems have been identified.

Following are some of the recommendations by experts on dealing with workload problems which are part of the job. Included are some widely recommended suggestions for staying mentally and physically healthy while dealing with workplace stress.

Recommendation one: limit hours

Studies are relatively consistent that burnout is a response to work overload and chronic time pressure problems (Maslach et al., 2001). Clearly, some organizations believe they can save and/or make money if they can increase workload. Workers, on the other hand, asked to give more both in the way of time and energy, sooner or later discover that an enormous conflict exists between their private, human needs and the needs of the organization.

In 2006, Columbia University economist Sylvia Ann Hewlett conducted a study on the prevalence of "extreme jobs," i.e., those in which individuals worked exceedingly long hours. Her summary of data collected on individuals working at a number of multi-national corporations revealed that

> 62% of high-earning individuals work more than 50 hours a week, 35% work more than 60 hours a week, and 10% work more than 80 hours a week. Add in a typical one-hour commute, and a 60-hour workweek translates into leaving the house at 7 am and getting home at 9 pm five days a week. If we focus on the subset of those workers who hold what we consider extreme jobs (a designation based on responsibilities and other attributes beyond pay), the hours are even more punishing. The majority of them (56%) work 70 hours or more a week, and 9% work 100 hours or more. (Hewlett and Luce, 2006, p. 51)

Not surprisingly, Hewlett discovered that half of the top group of executives she studied came home so tired that they were incapable of interaction or conversation with their families at the end of the day.

High-caliber professionals, regardless of age, gender, job sector, or continent, are working longer hours, according to Hewlett. But one does not need to work in the highest echelons of the business world to understand that the workplace is changing. Lower-ranking jobs not only demand more time, but are often more complex than they once were. Workers are also expected to fill multiple roles (i.e., multi-task) more often than they ever did.

The result is that it has become less and less possible to "turn off" the workday during personal time. Individuals headed for burnout not only work long hours, but often find themselves unable to stop thinking about work during their time away.

Pines and Aronson (1988) note that, although organizations are largely to blame, at least some of the fault may reside with the employee. They write,

> . . . people must learn to make a clear discrimination between the concrete demands of the job and the demands they place on themselves that they sometimes erroneously attribute to their "supervisor" or their "organization." Thus, some people regularly overwork, assuming that this is a demand placed on them by their organization. But if they examine the situation closely, they would realize that they were much harsher taskmasters than their employer. They would then realize that they *did* have more control than they

> realized – and would have to deal with the issue of whether or not they wanted to exercise that control. (pp. 30–31)

These authors urge employees to be severe about keeping the spheres of work and home separate. This means that the employee should try to be fully at work while at work, and then fully at home while at home. The strict separation of these two areas, if it can be accomplished, is less likely to lead to burnout.

Most important, however, is this: employees need to understand that a backbreaking workload is unsustainable in the long run. This is why employees working long hours must attempt to exercise some control and to make a commitment to themselves to reduce those hours by some amount, even if only by 15 minutes a day or one hour a week. A commitment to reducing hours has a double effect: it makes workers more cognizant of the time they are committing to the job and it starts the process of beginning to slow down the pace over time. Cutting back or cutting down on hours worked is the first major step one must take to avoid burnout.

Recommendation two: learn to say no

Individuals headed for burnout are often idealistic, enthusiastic and reluctant to turn down any new work assignments. Taking on too much work, however, results not only in exhaustion but also in certain tasks being given less attention than they require simply because an employee has no more time to give. As a result, employees feel they are doing a job which is less than adequate and the sense of themselves as competent workers suffers.

Even though some workplaces don't give employees much room to say no, there is often some wiggle room at the margins. For professionals, that might mean cutting back on volunteering for committees or doing less work outside the job with professional associations. Also, if there isn't enough time in which to do a proper job, the employee needs to say so up front.

But how you say it is as important as why and when you say it. Vaccaro (1998) discusses several ways to say no, and emphasizes that being polite, firm and honest is key. The "pleasant no," for example, is one in which you say pleasant things about the project and the person asking you to volunteer but then say no. She gives as an example the following: "[that project] sounds a lot more fun than what I'm going to be doing at that time, but I'm going to have to say 'no.' Thanks."

Another option for Vaccaro (1998) is a "conditional no," i.e., to offer to do a very small piece of the project while still saying no to the larger project. For example, an employee can explain that she can't be at the meetings, but might be willing to do some research or look into some small matter which is part of the project, leaving the rest to others involved. With the conditional no, it is vital to set boundaries and make sure that people know what it is you will and will not do.

With the "sleep-on-it no," Vaccaro suggests an employee can explain that she would like to think about it and will get back to the person making the request the next day. This avoids being put on the spot. It also gives the employee some chance to think about the pros and cons of saying no. As for the person who is asking you to do more work, it shows that even if you do say no, you have given it some serious thought.

The "alternative solution no," according to Vaccaro, gives the person looking to download more work another place to look for help. For example, an employee might say, "I can't do it, but I know Emily might not mind getting involved." It's good to know in advance, though, that Emily or whomever else you name won't resent your recommendation.

Finally, Vaccaro's "secret-weapon no" is simply to say no without explaining. The individual can simply say, "It's not possible for me to do that at the moment" and resist any efforts to explain further. The secret weapon, of course, has to be used judiciously; like all secret weapons, it needs to be reserved for special occasions and is not meant for daily use.

Of course, as noted above, sometimes no is an answer that will not be tolerated in the workplace. In cases where no is not an option, another possibility is to negotiate. The negotiation can be for more time in which to do the work, some assistance from others in getting the project done, doing the project on a smaller scale, or changing the parameters of the project. Also, if by taking on the job the employee won't have the time to do other important tasks, then the supervisor needs to know and to set the priorities. It is the supervisor who should decide which tasks can be left undone if the employee must say yes to this particular task.

Recommendation three: delegate

High achievers are unwilling to delegate, not only because they feel they can handle one more task, but also because they think they can do a better job than others even at the most mundane tasks. Management consultant Bonnie Elliot notes that employees who are unwilling to delegate don't know how to separate the wheat from the chaff:

> I find there's a common underlying assumption that [these workers] have to be involved in everything, because they fear that if they delegated management tasks, someone else wouldn't be able to do them as well . . . They wear it as a badge of honour that they have it all in their head, but I point out that's one of the reasons they don't have the time or attention to focus on their work. (Immen, 2012)

Nonetheless, Elliot also notes that, when questioned, overburdened employees often don't delegate less important tasks which fill their day because these are the very tasks they find most enjoyable. The key, she notes, is for the non-delegator to determine which, among the less important tasks, are the most enjoyable, and to keep doing those. Once these are identified, there should be some attempt to delegate or cut down on other work which is less enjoyable and which is preventing workers from getting some of their more important tasks done.

Recommendation four: manage time wisely

Individuals headed for burnout often have difficulty managing their time and have trouble cutting down on distractions, particularly from electronic devices. What follows are suggestions from a variety of experts in the field for getting the most out of each working day, and for thinking seriously about how time should be managed during the course of one's career.

a) Prioritize the work week

Employees need to monitor their work day to determine which hours are less busy and more suitable for certain types of tasks. So, for example, if it is easier to concentrate on difficult tasks (e.g. writing reports, working on budgets) when the office is quiet, then workers should try to schedule doing those difficult tasks for that optimal time. If that happens to be first thing in the morning, then more mundane tasks such as returning phone calls and checking email should be left for later in the day. Setting the phone so that it goes to voicemail directly and turning off email is crucial in order to make the most of this time.

Where long hours are a problem, the key is to make a list of most important to least important jobs in descending order and to stick to it. This technique involves setting goals for the work day and work week. Workers need to identify what must be accomplished immediately and what needs to be accomplished longer term. Of course, longer term projects can't be ignored and they will remain an acute source of stress unless a small amount of time is set aside for working on them throughout the week. By no means should longer term projects be left entirely to the last minute.

Once work has been prioritized in this way, pressing tasks should be addressed first, followed by a small amount of work each week on long term projects. If any time is left over, the less important tasks near the bottom of the list should be tackled.

Along the same lines, time should be scheduled not only for tasks and projects but also for higher level activities such as reflection or strategic thinking. Some time, no matter how small, should be built into the work week for these activities. Without scheduling, time for strategic thinking is something that is likely to be given short shrift or to fall between the cracks entirely.

No scheduling is required for daydreaming, however, and experts remind us how important it is to let your mind wander from time to time:

> "[p]rolonged periods of intense concentration can fatigue your brain," says Eric Klinger, Ph.D., a psychology professor emeritus at the University of Minnesota, Morris. "Allowing yourself a couple of minutes to daydream can provide just enough rest for you to regain your focus and be more creative and productive in the long run." The next time your mind takes a midafternoon meander, indulge your fantasies, like where you might go on your next vacation or what you would do if you won the lottery. (Wiener, 2007)

Scheduling or just finding a few minutes for some quiet time, whether it is done at home or at work, is also important, even if that quiet time is only a few minutes each day of deep breathing, meditation or staring out the window.

b) Minimize electronic interruptions

Constantly checking, monitoring and responding to email is a major stressor on the job, so much so that it even affects heart rate. Research has shown that employees who constantly check their email on the job tend to have 'high alert' heart rates which have been linked to higher levels of cortisol (a hormone associated with stress) in the body (UCIrvine Today, 2012).

A recent study found that cutting out email not only resulted in better focus for workers but it also changed their heart rates. Employees who were not permitted to check their email for five days were found to have more natural and variable heart rates. They were also less distracted.

Researchers found that these employees switched between computer screens half as much as their email distracted counterparts – from 37 times an hour to 18 times an hour (UCIrvine Today, 2012) – resulting in better focus on the task at hand. As a result, limiting access to email or checking email only at certain times during the day is one concrete way to reduce stress.

c) Schedule family time and downtime

Scheduling family time and downtime, and setting boundaries between work and home life, is enormously important to keeping burnout at bay. Researchers have found that people without children tend to burn out more often than people with children. Part of the reason has to do with investing emotional and physical resources disproportionately in the workplace when you are not forced to manage your time in and around family (Senior, 2006). Another reason has to do with the emotional support that family provides.

Individuals at risk for burnout need to make an effort to re-establish proportionality between work and home. One way to do this is to decide to eat at least one meal a day with friends and family rather than with work colleagues and to limit making work calls from home. Also, just as work commitments are pencilled into the calendar, family commitments need a place there as well. As one commentator notes, "Keep the commitment once made and expect to experience withdrawal symptoms and guilt [about work]" (Kaufman, 2005). Once the commitment becomes more of a routine, the guilt should subside.

Downtime means taking time for yourself by scheduling time alone and away from electronic devices and cell phones. This time should be devoted to activities such as hobbies, reading for fun (instead of for work), taking a bath, or getting a massage. It can also mean time for sitting quietly in a garden and day dreaming.

d) Periodically re-evaluate commitments

Sometimes employees find it difficult to say no to more work because they are not certain what they want from their job or clear about where they hope the job will eventually take them. Before agreeing to additional work, employees should ask themselves what the benefit might be in saying yes. Will it lead to greater control and autonomy, for example, or to a better position? Will it make the job less boring or difficult? Will it allow for the opportunity to network with others who might help advance their career in the future? Saying no is easier when it's clear what important factors should be in place in order to say yes.

Baird and Baird (2005) note that professional or workplace commitments which do not add to any of one's long term career goals should be abandoned. They also suggest re-evaluating and prioritizing commitments depending on the stage one is at in one's career. For example, new employees are more likely to volunteer for certain activities or opportunities in order to build a career. Once that career is established, however, it might be time to re-examine those commitments and start paring them down so that only the most important – those necessary to sustain a career – remain.

Recommendation five: take vacations

The benefits of vacations on one's physical and mental well-being are real and yet individuals are frequently reluctant to take them, often because of workload issues or because taking a vacation might be perceived negatively by the employer. A survey in the UK by the travel reservation website Expedia conducted in 2010 found that 20 percent of those surveyed do not take their annual leave because of work pressure (expedia.co.uk, 2010) and half of those who do take vacation feel compelled to check email and phone calls while on vacation (Expedia.com, 2011).

Americans are not much better. On average, workers in the US were allotted 14 vacation days a year but only used 12 (Expedia.com, 2011). At the end of 2011, approximately 57 percent of working Americans had not used up their allotted vacation. The most common reason given was that they had too much work (Censky, 2012).

Sixty-two percent of Canadian workers reported that they continued to check email and voicemail while on vacation (Expedia.ca, 2011). In addition, Canadian workers tend to leave one day of vacation on the table, taking an average of 15 of their 16 allotted vacation days. This leaves a total of 17 million unused vacation days overall in Canada (Expedia.ca, 2011). Thirty-eight per cent of Canadian workers have cancelled or postponed their vacation because of work commitments.

Not surprisingly, research has found that levels of stress drop significantly immediately after a return from a vacation (Etzion, 2003). People are more active on vacation, they tend to sleep better, and have the time to rejuvenate mentally. Lower levels of stress have been found in employees who took longer vacations compared to those employees who took shorter vacations; nonetheless, the ameliorative effect on burnout was found to exist whether vacations were short or long (Etzion, 2003). Rates of burnout dropped for approximately one month before rising again. As a result, several short breaks throughout the year (for example three one week vacations as opposed to one three week vacation) are thought to achieve the maximum, positive long-term effect in ameliorating stress (Etzion, 2003).

As important as the lowering of stress levels, time away from the office also gives an employee some distance. As mentioned earlier, employees often don't notice or don't understand the signs of burnout. Getting away from the workplace gives the employee the time and space necessary to analyze why he or she is feeling stressed. Having the time to mull it over might also lead to ideas on how better to cope with the situation.

Recommendation six: break the routine

Boredom is as stressful as too much work and some experts believe that, as workplaces are becoming more boring, workers are becoming less willing to tolerate boredom (Hoare, 2012). Studies have shown that even though the busiest employees are the happiest (Joyce, 2005), it is difficult for organizations to find a balance between the type of "busy" which is just right, versus work which is overwhelming or underwhelming. This shows in the statistics: 55 percent of all US employees do not feel engaged at work or psychologically connected to the organization (Joyce, 2005), and one in four British workers reports feeling chronically bored ("25 percent bored," 2012).

Despite numbers which indicate a relatively wide scale boredom problem, organizations are generally unwilling to address it because of the signal it sends about the institution and the work itself. As a result, when the organization is unwilling to deal with boredom, it is up to an employee to find some way to shake up the routine at work. Even making small changes in the work day is important for both mental and physical well-being.

For example, incorporating new activities into the work day or carrying out old activities in different ways or at different times is a start. Employees who eat lunch at their desk each day while trying to work not only have no chance to unwind, but they risk the physical problems which result when eating while under stress. As Ashley Koff, RD, a Los Angeles dietitian, notes,

> [s]carfing down your sandwich may help you plow through that to-do list, but it certainly won't help you feel calmer . . . "When you're stressed out, your body diverts its resources to release hormones that help you cope, decreasing the production of the hormones

> associated with digestion," she explains. So if you eat under duress, you'll likely experience indigestion or bloating, which only makes you feel worse. "A half-hour break may not be realistic on a busy day, but you can close your door, turn off your e-mail, and concentrate on eating your food one bite at a time," Koff suggests. (Wiener, 2007)

Another way of combating stress is having lunch away from the office, eating lunch outside or walking at lunch.

Experts also recommend identifying a hobby or a pastime which you feel passionate about and trying to engage in that activity over the lunch hour:

> [w]orking on your hobby puts you in a zone, allowing you to forget about other stressors, which is especially useful at the office . . . If you're a foodie, use your lunch break to browse new recipes online; knitters might tote their needles and yarn to the office cafeteria. (Wiener, 2007)

Along the same lines, organizing a lunch-hour activity among co-workers who are interested in a specific pursuit, or becoming involved in an activity already underway is another way to lessen stress.

Breaking up the daily routine by getting up at a different time, taking a different way to work or a different way home can also help. Re-organizing work space is yet one more way of inserting some change into the daily routine (Elson, 1979).

Finally, in order to break up the routine, Pines and Aronson (1988) suggest that individuals who work in cerebral, verbal jobs should look to incorporate more physical and non-verbal activities into their off-hours, and vice versa. For workers who sit at a computer each day, more walking, biking and spending time outdoors should be incorporated into time away from work. In contrast, workers who do physical work all day are better to read, listen to music, and find other ways of incorporating more cerebral hobbies during their time away from the job.

Recommendation seven: increase resilience

Shawn M. Talbott in *The Secret of Vigor* (2012) explains that eating wisely, taking dietary supplements, getting enough sleep, employing stress management techniques and exercising are necessary in order to increase resilience and combat stress and burnout. Talbott calls these

"Vigor Improvement Practices" (VIPs) and emphasizes the importance of each of these in maintaining both mental and physical health. Talbott's VIPs are discussed below.

a) Eat healthy foods

Employees at risk of burnout often neglect their physical health and this includes eating healthy foods. Constant stress is hard on the immune system which also relies on a nutritious diet for its proper functioning. A healthy diet is one way to combat stress, and food "can either be the safest and most powerful form of medicine, or the slowest form of poison" (Steckler, 2012). This means that eating properly can prevent, cut down or eliminate a wide range of health problems including obesity, hypertension, inflammation, diabetes, heart disease, sleep and mood problems, etc. In contrast, a poor diet will stress the body, and lead to greater susceptibility to disease as well as poor concentration, mental exhaustion and fatigue.

Individuals who work in chronic stress environments often have difficulty eating healthily. They might skip meals altogether or conversely use food as a way of reducing stress. Comfort foods which are high in calories, fats and sugars trigger the release of certain chemicals in the brain which results in feelings of pleasure and a desire to eat more of the same types of foods. Done too often, the comfort food results in additional weight gain turning pleasure into pain as it adds to the stresses placed on the body.

Burnout has been shown to play a significant role when it comes to problematic eating patterns. Recent studies have shown that women who burn out are prone to both emotional eating and uncontrolled eating (Nevanpera et al., 2012). Emotional eating occurs when an individual is stressed or unhappy and eats not because she is hungry but because she wants to alleviate those feelings. Uncontrolled eating, in contrast, occurs when an individual constantly feels hungry and can't stop eating until the food at hand has been consumed. Both are serious problems in that individuals who are stressed are more likely to eat food that is bad for them rather than binge uncontrollably on healthy food such as fruit or vegetables.

Stressed employees tend to neglect their own health and that includes forgetting to eat regular meals consisting of a variety of healthy foods including proteins (milk products, eggs, fish, meat and meat alternatives), vegetables, fruit and whole grain products such as whole grain bread, oatmeal, brown or wild rice, and whole wheat pasta. How one eats is also important. Throwing back a meal in the car or at the desk is more likely to result in overeating or eating unhealthily since no attention is being paid to the activity.

Finally, experts recommend a daily multi-vitamin to supplement the nutrition that is not being ingested when workers fail to eat the recommended number of servings of fruits, vegetables and other foods each day (Talbott, 2012). Certain vitamins or minerals prevent diseases or carry out other functions in the body not performed by any other vitamin or mineral. Missing out on any one of these can mean missing out on a vital health benefit.

b) Get enough sleep

Sleep has a restorative function and proper sleep is necessary for, among other things, tissue renewal, proper brain functioning including concentration and mood, and a healthy immune system. A lack of sleep has been linked to obesity (Brody, 2011) and to premature death (Smith, 2010). Researchers who have studied the effects of lack of sleep on levels of attractiveness have found that sleep-deprived individuals are perceived as being less attractive and healthy than their sleep-refreshed counterparts (Brody, 2011).

In addition to the problems associated with health and physical appearance, not getting enough sleep has also been shown to predict clinical burnout (Söderström et al., 2012) and may be linked to the excessive fatigue which is a hallmark of burnout. If this is the case, researchers comment, it could mean that burnout has a confirmed physiological as well as a mental component (Ekstedt, 2005).

c) Exercise

Not only is exercise a reliable way to relieve tension and stress, but researchers have discovered that employees who exercise are less likely to suffer from burnout and depression (Toker and Biron, 2012). Individuals who exercised four hours a week were found to be half as likely to suffer from job-related mental health problems as those who did not exercise at all.

As for the best type of exercise to combat stress, the American Running Association recommends that

> using large muscle groups in a rhythmic, repetitive fashion works best. This type of movement seems to foster a kind of muscular meditation. Running is a prime example, but even walking for 20 minutes can clear the mind and reduce stress. Vigorous workouts like elliptical training are even more popular stress reducers, perhaps

> because you gain a sense of "taking it out on" something, yet in a productive way that ultimately contributes to your self-image. (American Running Association, 2011)

In addition to running or walking, experts add swimming, cycling, tai chi and other aerobic exercises to the list. Exercise helps the body release the feel-good chemicals endorphins and serotonin which help to reduce stress and to lighten mood. Regular exercise also makes people feel more energetic and helps them to sleep better.

d) Employ relaxation techniques

Relaxing is an important way of dealing with stress. Relaxation is more than merely putting one's feet up; in order to get the benefits, specific relaxation techniques such as deep breathing, muscle relaxation, meditation, and yoga need to be employed to bring about mental and physical changes which calm the mind and help fight stress.

The reason why simply sitting down and putting one's feet up is not good enough is that in order to be effective, a relaxation technique needs to "break the train of everyday thought, and decrease the activity of the sympathetic nervous system" (Goleman, 1986). The sympathetic nervous system is responsible for the "fight or flight response" which occurs when the body secretes hormones necessary to help the body either challenge or run away from a perceived threat.

Employees don't need to be facing an actual "fight or flight" situation in order to have their bodies respond the same way. Simply working in a stressful environment can overwork the adrenal glands and leave them no longer functioning optimally. Adrenal fatigue is characterized by a number of symptoms including inability to sleep well, excessive fatigue, irritability, craving excessive salt or sweets, poor concentration, low stamina, low blood pressure, lower resistance to colds and other health problems, etc. Luckily, adrenal fatigue can be helped by eating a healthy diet, taking vitamins and practicing relaxation techniques to alleviate stress and bring about a physiological state called the "relaxation response" (Goleman, 1986).

The relaxation response is responsible for changes in hormone levels that help the immune system fight infectious disease. Relaxation releases muscle tension, lowers blood pressure and slows the heart and breathing rates. Research has also discovered that relaxation techniques improve blood flow to the heart, lower cholesterol levels, improve the body's ability to regulate glucose, and alleviate chronic pain (Goleman, 1986).

Job domain 2: control

We move now to the second of Leiter and Maslach's (2005) job domains – control – and to what an individual employee might do to deal with a control problem in the workplace. Control has to do with how much say workers have over the job, including such matters as how the work is done, the hours during which one does the work, and even when one is permitted to take breaks or vacation. Research is clear that the amount of control workers have over various aspects of their job is important in determining whether they might be at risk for burnout. In this section, we discuss two aspects of control: autonomy and supervision.

a) Autonomy

Autonomy problems exist when workers have little or no control over how the job is to be done, or when their point of view seems not to be valued by management. Additional control issues include being micromanaged, having no scope for creativity, or dealing with weak leadership or ineffective teams. Issues also include having responsibility for a job over which one has little authority or decision-making power. As one burned out worker noted,

> I started having less and less faith in the direction the senior management was taking the organization, but they didn't want to hear anything that differed from their own view. I felt like I had the responsibility of being on a team without any possibility of having any input. . . . It was always clear to me that I did not lose faith in the work. I lost faith in the work relationships. And in my ability to make any difference in them. (Glouberman, 2002, pp. 30–31)

Lack of autonomy, studies tell us, is responsible for higher rates of burnout (Aiken et al., 2002; Verissimo et al., 2003; Keeton et al., 2007; Day et al., 2009). Not surprisingly, workers who are given greater autonomy and the ability to make choices about how the work is done have been shown to have better morale (Ward and Cowman, 2007), greater job satisfaction, more commitment to the organization, greater involvement in the job, better performance, fewer mental and physical problems, less absenteeism and less turnover (Spector, 1986).

The problem for most workers is how to gain more autonomy in an environment which offers little. Leiter and Maslach (2005) suggest identifying the type of control which is absent in the environment and then responding accordingly. For example, in an environment in which the employee is micromanaged, they suggest micromanaging back; i.e., shaping the flow of information that returns to the supervisor in order to better steer responses and outcomes. They also suggest testing the limits of control by starting small and determining whether acting with more authority at the edges of the job might lead to a loosening of oversight.

Another way to increase job autonomy is by building a record of exceptional performance on the job and earning the trust of management. This, of course, is something which can take some time to accomplish.

Volunteering to take more control is also an option. Workers who are aware that their supervisors are themselves overwhelmed on the job can suggest that some control be shifted their way. In this type of situation, offering to take on more responsibility might be a welcome suggestion to a supervisor struggling to tread water.

Finally, the way an employee might achieve more autonomy in the workplace is through external validation. Volunteering for professional associations, and receiving kudos through these associations, is useful in shifting the balance of power in the workplace as managers are often unable to ignore the opinions of their peers in other institutions (Leiter and Maslach, 2005).

b) Supervision

Weak or inefficient supervision is considered a control issue (Leiter and Maslach, 2005) and has also been associated with increased job dissatisfaction and higher levels of burnout (Hyrkas, 2005). Leiter and Maslach (2005) suggest ways to deal with weak or inefficient leadership by taking creative control, critical control or supportive control.

Creative control involves keeping the boss up to date on innovation, new ideas or new concepts and working with the boss to make both the monitoring and the implementation of new ideas part of the job. Supervisors may have little time to keep an eye on what is new in the field and offering to take on this role may serve to relieve some of the work burden of management and shift some control to the employee.

Critical control involves taking more interest in job performance and working with the supervisor to enhance quality in the workplace. This is a somewhat trickier role as employees who monitor the work or output of others quickly find themselves unpopular. Quality and performance are often best discussed in groups or teams where everyone has a voice and can offer suggestions. Employees interested in critical control might offer to lead or facilitate this type of group or team. Seeking to improve the quality of the work democratically is more likely to be acceptable to the group as a whole and to be less divisive in the workplace.

Finally, supportive control involves paying more attention to the supervisor and communicating more about frustrations and accomplishments in an attempt to model supportive behavior. The hope is that it will eventually be returned in kind (Leiter and Maslach, 2005). Supportive control is about being a shoulder to cry on or a safe person in whom to confide. The theory is that supervisors who find social support in their employees are more likely to return that support when the employees themselves are in need of assistance.

Clearly, how one responds to a lack of control depends on the specific problem encountered on the job. But these suggestions remind us that, even in a job which appears to have no room for autonomy whatsoever, there might be a small amount of room at the margins for ways to begin to improve the situation.

Job domain 3: reward

This is the third of Leiter and Maslach's (2005) six job domains, and these authors offer several suggestions for employees whose job rewards are insufficient. Reward, as we are reminded, is often equated with money; however, it can also mean a great deal more including more recognition, greater status, or work which is more satisfying.

At its most basic level, trying to fix the problem of insufficient compensation is not always easy; many organizations have pay charts which prevent employees from negotiating any salary other than the one they have. Also, supervisors in larger institutions have no authority to re-negotiate salaries once an employee has been hired. On the other hand, where salary is negotiable, Leiter and Maslach (2005) suggest doing one's homework, coming equipped with empirical data and discussing the topic dispassionately. It's also good to have a Plan B; where money discussions are off the table, perhaps something like more time away from the office, better hours or an improved physical environment might be attainable.

It is, of course, possible that the organization might be unwilling to commit to non-monetary rewards and remain firm about no additional compensation. Where that is the case, these experts recommend reducing output so that it is more in line with the salary. Interestingly, research has shown that workers seem to do this as a matter of course when salary and workload are seriously mismatched. It seems we naturally do less when the compensation for the job is thought to be correspondingly less than it ought to be.

Sometimes money is not the issue and what workers are missing is a pat on the back. Where acknowledgement is the problem, these authors suggest discussing the situation with one's supervisor and asking for more feedback on the job. Another possible suggestion could be bringing various accomplishments to the attention of the supervisor. If the supervisor seems to be slow in getting the point, these authors suggest that workers try, in the meantime, to reward themselves. This could involve bringing accomplishments to the attention of others in the workplace, or finding some private way to celebrate accomplishments. Employees who feel undervalued might also want to seek recognition beyond the workplace such as working in the community or with professional associations.

Where better job assignments are the desired reward, Leiter and Maslach (2005) suggest either negotiating better job assignments or working on expanding or increasing the parts of a job that are enjoyable. Changing the way the job is done might also lead to greater satisfaction.

Job domain 4: community

This is the fourth of Leiter and Maslach's (2005) six job domains. When problems arise in the area of community, it often means co-workers can't get along, supervisors are disrespectful or undermining, subordinates are angry and resentful and clients are problematic. Healthy communities are those in which people can easily talk to each other, will co-operate with one another, and will assist one another in solving problems. Not surprisingly, then, social support in a number of studies has been found to act as a buffer against burnout.

Conflict in the workplace has been linked to burnout, whether it involves supervisors, co-workers or others. For example, the matter of support from co-workers has been found to contribute to predictions of burnout (Greenglass et al., 1997); in other words, the greater the support, the less likelihood there is of being at risk for burnout. In contrast, co-workers who are competitive, distant, back-biting, indifferent and/or unpleasant clearly contribute to a more stressful environment.

Along the same lines, the population group with which a professional works can be a source of stress. Conflict with clients, for example, has been found to be significantly associated with emotional exhaustion and depersonalization (Fujiwara et al., 2003), as well as a major cause of burnout.

In terms of management, poor supervision has also been found to be not only one of the leading causes of burnout but also responsible for low productivity ("Much worker burnout," 1992). Where subordinates perceive their supervisor to be abusive, for example, they report greater emotional exhaustion (Tepper, 2000) and depersonalization (Yagil, 2006). In contrast, when supervision is supportive, supervisors can actually provide a buffer against stress. Supportive supervision means that supervisors communicate with workers, provide constructive and encouraging feedback, and are involved with their employees on a regular basis rather than just when a crisis looms (Pines and Aronson, 1988).

Where the workplace "mismatch" for an employee is one involving community, Leiter and Maslach (2005) recommend as a first step determining who the problems are – co-workers, supervisors and/or clients – and where the solution might lie. If there is serious conflict in the workplace, it must be addressed in some way, either by using conflict resolution techniques available through the workplace or by trying to set something up through management.

Ultimately, building a better community starts with better communication and by reaching out to others in the workplace. Sometimes, however, communication itself is the problem. In cases such as these, staff meetings are one area in which poor communication might be remedied. Meetings are a place in which shared goals can be discussed, and employees should feel safe enough to bring complaints, let off steam, and discuss problems. Pines and Aronson (1988) suggest establishing rules for these meetings in order to ensure that they do not become one more area in which employees do battle. They see meetings as a place in which complaints and problems can be aired, but solutions must be explored without criticism, sarcasm or judgment. Once this has been done, the best solution, they recommend, should be chosen from among those suggested. Meetings such as these let staff members feel that they have a voice and some control over problematic situations, and that it is the job of the collective not only to air grievances but also to work towards solutions together.

Where there is little conflict, but also little cohesiveness, employees need to work on building more collegiality and team spirit. This can be done through social activities and group events which can be organized and built somewhere into the work week. Leiter and Maslach (2005) suggest reaching out to others in the organization, organizing a support group or a community service project. Both asking for and giving support are yet other suggestions to try to build better relationships in the workplace and to create alliances.

Job domain 5: fairness

This is the fifth of Leiter and Maslach's (2005) six job domains. Fairness in the workplace has to do with treatment which is even-handed, non-discriminatory, and does not favor certain individuals over others except when it is merit-based. As Maslach and Leiter (1997) note, "[a] workplace is perceived to be fair when three key elements are present: trust, openness, and respect" (p. 52).

Perceptions of fairness are important for more than contributing to job satisfaction – bad bosses also have a significant effect on physical health. Workers who reported their supervisors as acting unfairly were found, in one study, to have "a 16 per cent increased risk of heart disease and a 38 per cent increased risk of stroke" (Naish, 2005). In contrast, those employees who felt that their supervisors acted fairly were found to have one-third less heart disease than those who rated their managers as unfair or neutral (Naish, 2005).

Fairness involves equitable treatment; it also involves respect. Leiter and Maslach (2005) note that promoting respect in the workplace means promoting civility. One of the first ways to address incidents of disrespect or discrimination is to talk about the incident with the offending party, or when that does not change the behavior, to talk to a supervisor. Where this fails to resolve the problem, the aggrieved employee should consider making a formal complaint through the organization's human resources or equity officer.

Where a formal complaint fails to resolve the problem, the employee should then consider taking the complaint to an external organization or government body. Taking legal action is also an option. These are more drastic steps and not all organizations welcome them. The response will differ depending on the organization – anything from complete assistance to full-blown reprisal. An employee looking to take a complaint to a human rights body or to the court should first seek advice in order to consider the various options. Leiter and Maslach (2005) counsel that employees should weigh the risks of going outside the organization and understand the consequences.

Job domain 6: values

The sixth of Leiter and Maslach's (2005) job domains involves values. Work values are an employee's perspectives or standards which help guide him or her on the evaluation of what is right and wrong in the workplace. Mismatches in values occur when employees are required to choose between the values of the workplace and their own personal values. An employer might ask an employee to do something which he or she feels is unethical, even to tell lies on behalf of the organization. For example, employees might be asked to explain to clients that a change in service is for their benefit when, in actuality, the change is only being made to save the organization money.

A mismatch of values is relatively common; a study published in 2007 found that 30 percent of employees surveyed said that their employers' values were not always aligned with their own ("30 percent of workers' values," 2007). Leiter and Maslach (2005) note that when mismatches occur between the values of the organization and the values of the individual, the first thing an individual must do is to try to pin down the specific problem. Employees might, for example, perceive the organization itself or the supervisors within the organization for which they work to be destructive or dishonest. Somewhat less dramatically, employees might feel that, even though their organizations are engaged in honest and ethical pursuits, their own individual jobs or tasks within the organization are meaningless. Each problem is significant, but each calls for a different response.

Employees looking to deal with problems of dishonest behavior in their employing organizations have several options. The first is pre-emptive; e.g., they might approach management to request training on ethical decision-making for their group and then take advantage of any discussion that arises from that training to suggest changes in work practices. But where subtlety is ineffective, employees can attempt to change the environment by going to a figure in authority and blowing the whistle. Whistle-blowing requires careful preparation; organizations often respond defensively either by attempting to downplay the problem or by punishing the whistleblower himself. Termination of the whistle-blower's job is, unfortunately, not a rare occurrence.

Employees must prepare carefully if they plan to shed light on illegal or unethical activity. One of the most important things is choosing the right individual(s) to contact. It is also vital that employees have enough evidence before making an accusation (Leiter and Maslach, 2005). Sometimes, there may be no fix for an employee who is involved with an organization that is involved in seriously unethical or illegal activities. Leiter and Maslach (2005) note that when all else fails, a serious mismatch in values might mean looking for a new job.

Where the organization is involved in practices which an employee feels are destructive but change does not appear hopeless, Leiter and

Maslach (2005) recommend taking steps from within to change the organization's values. This could involve volunteering for committees or working on projects which give an employee a chance to voice his concerns from within the organization. Changing the values of an organization involves promoting discussion on organizational impact, responsibility and constructive initiatives. Working from within the organization might prove slow and frustrating, but keeping at it could eventually lead to constructive change.

When working within the organization fails to pay any dividends, working outside with professional associations, charities or volunteer groups might help to make a difference to the individual. So, for example, in the case of a library which has new policies designed to make it more difficult for homeless individuals to spend much time there, an employee who is unable to change the policies from within might be interested in volunteering at a homeless shelter or soup kitchen in her spare time.

Sometimes, though, the problem is not a clash of values, but simply the feeling that the work is not meaningful. In a survey of job seekers in 21 different countries, 60 percent of the survey participants selected "challenging and interesting work" as the primary thing they were looking for in a job, followed by 58 percent who chose "recognition and reward for their accomplishments" (Beauchesne, 2006). "Meaningful," of course, has different significance for different people, but is nonetheless important. Employees who feel they have meaningful work, however that term is defined, are likelier to be healthier, happier at home, and to have better relationships (Bruzzese, 2010).

To build meaning into a job, these authors recommend finding ways to take pride in the quality of the work or to focus on providing better service to clients. Employees can also add meaning outside of work by having the workplace become involved with charitable causes, or by finding volunteer work which has meaning.

What the employer can do: burnout prevention in six job domains

Introduction

Supervisors and their organizations have many reasons to prevent burnout in employees. For example, burned out employees suffer from an all-encompassing fatigue which manifests itself in a number of ways, including signs of boredom and depression. Employees who were once sociable now seem to deliberately withdraw from contact with others, while employees who were once withdrawn might now exhibit signs of anger towards others. As Glicken and Janka (1982) explain,

> [t]he burned out individual is smoldering in place, riddled with stress and dissatisfaction and responding with an ever intensifying loss of energy and interest. The burned-up individual has progressed to such a degree of lethargy and immobilization that he or she is generally destined for a radical event, such as mental or physical illness, sudden resignation, or involuntary termination. (p. 69)

Employers, unfortunately, have a tendency to view burned out employees as weak – unable to muster the physical and psychological strength needed to get over the bumps. These employees are seen as being difficult or suffering from some psychological disorder. Some organizations write off employee burnout as simply failing to keep up with the organization's high performance standards.

This approach is more transparent than organizations would like to admit. Employees in one study who were asked about organizational practices and the extent to which they felt cared for as they burned out noted that

> their organizations made the right noises about ensuring people didn't burn out, but in practice tended to take a passive, reactive stance on work/life balance; they acted in a crisis or when asked to do so. Some mentioned that their employer had set up systems to try and prevent burnout but few would be willing to use them because of an endemic lack of trust in the organization's motives. Indeed, in some organizations, there was a stigma attached to seeking support that was available. Overall, most felt that their organization saw burnout as a sign of weakness or failure. (Casserley and Megginson, 2009, p. 41)

This is, of course, a problem for organizations. To see burnout merely as the result of personal shortcomings or personality is to risk losing employees who were once high achievers and who now either fail to perform adequately or attempt to cope with their symptoms in ways which are personally and professionally destructive.

Instead, the savvy workplace will build in prevention programs, intervention assistance and treatment programs to shield employees against burnout. Prevention programs such as stress management training courses and information on stress management are aimed at workers generally. In contrast, stress management intervention and treatment programs are often tailored to the needs of specific individuals. In the case of stress management intervention, managers can be trained to keep an eye out for behavior that is unusual from their employees including reduced productivity, absenteeism and increasing self-isolation. Managers can have a private talk with the employee to find out what is going on, and can then recommend help if that seems like the appropriate next step.

Organizational fit

Managers should have some understanding of the personal characteristics of the employees in a particular role and whether there is a "fit" between these characteristics and the demands of the job. This can be accomplished through career development programs which help individuals understand their own strengths and weaknesses on the job. Fit can also be identified in terms of how well an employee is meeting the demands of a particular job and seems to be aligned or comfortable with the values and goals of the workplace.

There is more to job–person fit, however, than simply having an accurate understanding of what the job entails. Maslach et al. (2001) note that organizations often think of "fit" simply in terms of hiring the right individual or seeing that that individual adjusts to a new job. Burnout, however, is not a problem which affects only newcomers; instead, it is as, or more, likely to affect individuals who have been doing a particular job for some time and who are chronically mismatched with the job. In this context, "fit" is not merely the match between an employee's skills and the tasks involved in a particular position; it also takes into account an employee's emotion, motivation and response to stress, along with the larger organizational context.

Given this broader and more complex view of job–person fit, these authors have considered six specific job domains which we discussed above: workload, control, reward, community, fairness, and values (Leiter and Maslach, 2005). Each of these areas provides fertile ground for dissatisfaction with one's worklife. We have discussed these areas from the standpoint of the employee and what that employee could do when fit in one or more of these areas was a significant problem. Now we turn to an examination of the same six job domains, but this time from the standpoint of organizations. The following are recommendations from Leiter and Maslach (2005) as well as other experts and researchers who offer advice on what organizations can do to reduce burnout in their employees.

Job domain 1: workload

Recommendation one: reduce excessive workload

The first of Leiter and Maslach's (2005) six job domains – workload – is a problem when workload is excessive, when time pressures are too severe, when work is too complex or when workers feel they must be on call constantly. Managers must take steps to deal with these matters or face personnel problems such as illness, absenteeism, reduced output and turnover further down the road.

When, for instance, workload is expected to be busier than usual, but only temporarily, managers need to let employees know what has brought about this change, how long the increased amount of work is expected, and why additional staff are not being hired. They should explain their expectations but also ask employees if there are any scheduling problems or potential conflicts such as family or other commitments that might make it more difficult or impossible to meet an increased workload. An effort should be made to help such employees meet their commitments by tailoring the workload and the schedule if possible so that the disruption to the employee is kept at a minimum.

Managers also need to lead by example. Where employees are expected to take on an excessive load for a special project, for example, managers should be doing the same (Tunick, 2000). Employees are unlikely to go the extra mile if they suspect that the pain is not being shared by their supervisors.

Recommendation two: identify and reduce boredom

While an excessively demanding workload can lead to burnout, an undemanding workload can be equally as stressful and bad for the organization as a whole. Indeed, in certain circumstances, even employees with demanding workloads can find themselves as bored as those with not enough work to do (Hoare, 2012).

The problem with boredom in all its forms, however, is that workplaces are particularly unwilling to admit that their workers might be bored.

Not to do so, however, is a problem. Bored employees harm their organizations in a number of ways: by picking on others, by deliberately failing at tasks or sabotaging work, by stealing, and horseplay. Bored employees also tend to be late or absent more often, or to take longer breaks (Hoare, 2012). They spend the day surfing the Internet, writing emails or taking care of personal business (Joyce, 2005). Studies have shown that workers who anticipate a day with a light workload are two and a half times more likely to call in sick ("Work boredom," 2012). Accordingly, even though busy employees might be bored, they are less likely to be prone to absenteeism.

One way to combat boredom, experts suggest, is to create "a culture of 'psychological safety' in which 'it's okay to ask questions' " (Hoare, 2012). Allowing employees to talk about the doubts and worries they have, along with their feelings on boredom, boosts morale. Organizations which can learn to be less defensive about this type of honesty will see better results in the long run.

Supervisors may also have it within their power to organize work such that employees are given some relief. This involves allowing workers to change tasks periodically, to rotate through various duties, or to experiment with new ways of accomplishing the same tasks. Providing workshops and additional training sessions, not only geared to the work itself but also to mental and physical health, helps to alleviate the stress and to provide guidance on coping. Shorter work shifts, allowing personnel to take more breaks, or providing opportunities for lateral job swaps also offer some variety in the workplace.

Recommendation three: model desired behavior

Managers need to be aware they model behavior including work–life balance. Where managers work excessive numbers of hours and do not take time out for family or vacations, they signal to subordinates that this is expected behavior and, in turn, increase the levels of stress in the workplace.

A stressed-out manager may exhibit signs of irritability, impatience, frustration or irrational thinking. Employees are less likely to be creative or candid with such a manager, and are less likely to communicate openly (Kaufman, 2005).

In contrast, when managers model healthy behavior, they improve working conditions for both their employees and themselves. One example is scheduling a "walking meeting" where meetings take place while going for a walk around the block. Walking meetings are healthy in that everyone is getting some exercise and fresh air; changing how meetings are done may also stimulate creativity. Managers also need to remember to take breaks from time to time in order to let employees know that it's okay to occasionally step away from their desks (Ray, 2012).

Recommendation four: don't skip vacations

Because excessive workload is consistently linked to burnout, steps should be taken to ensure that employees are taking vacations. Employees who take vacations are more likely to be physically and mentally healthier and more productive than their non-vacationing counterparts, but the organization must take some responsibility to ensure that employees do not forgo vacation (Dixon and Wills, 2010).

Among the steps that organizations can take is drafting a vacation policy which mandates that employees use their vacation time. The same expectation must be set for managers. Employees and managers should also be given guidance on how to schedule vacation time (i.e., scheduling in advance, scheduling around the vacations of other workers, and indicating the optimal amount of time, from the organization's perspective, to take for any vacation) (Dixon and Wills, 2010). At the same time, organizations should also be careful to give employees enough latitude in determining for themselves the dates and duration of their vacations so that they can work out scheduling around family and other personal commitments.

Recommendation five: consider sabbaticals

In 2012, one-quarter of the companies on Fortune's 100 Best Companies to Work For offered paid sabbaticals, and many more offered unpaid leave (CNNMoney.com, 2012). More companies are now offering sabbaticals as a recruiting and retention tool and to ward off burnout. Sabbaticals are primarily associated with academia in order to foster professional accomplishment. Outside of academia, however, sabbaticals are often taken for the purpose of pursuing education (job-related or not), additional training, research or for personal or family reasons.

For employees who are burning out, a sabbatical might be the first big step towards recovery. Ironically, individuals who are at risk for burning out might be those most reluctant to relinquish control of their specific area of work. Getting time away is crucial in terms of establishing some perspective about the job. Sabbaticals alleviate stress; employees who take them talk about coming back feeling relaxed and rejuvenated (Gardner, 2007).

Organizations which offer sabbaticals see them as a means of attracting, retaining and rewarding employees. In addition, both the organization and the employee benefit when the employee returns revitalized and with renewed enthusiasm for the job.

Sabbaticals are not a universal panacea, however. For one thing, not everyone can take advantage of the opportunity – in cases where sabbaticals are unpaid, some employees may not be able to afford forgoing remuneration. Others might have family commitments which prevent them from travelling or taking advantage of various opportunities.

Also, while sabbaticals help heal employees at risk for burnout, employees returning from sabbatical will face a similar risk upon return if changes have not been made to alleviate key stressors in the workplace (Giancola, 2006). What makes the sabbatical so ideal in this situation, however, is that time away is likely to help the employee pinpoint the specific problem or source of stress, allowing the employee to try to address it with the organization and to make changes upon her return.

Job domain 2: control

The second of Leiter and Maslach's (2005) six job domains – control – is one that is often difficult for organizations to share, and yet not sharing control with employees is one of the key ingredients of job strain. Job strain has been defined as "a measure of the balance between the psychological demands of a job and the amount of control or decision-making power it affords" (Wilkins and Beaudet, 1998, p. 47). Robert Karak's job-strain model posits that workers most at risk physically and mentally from stress are those who have high job demands (such as workload or time pressures) and little say in how the work should be done.

Policies in the workplace generally give workers little scope to change their jobs even if those changes would allow for improvements. Instead, workers are closely monitored and work is expected to be carried out in a "narrow, cookie-cutter, one-size fits all-approach" (Maslach and Leiter, 1997, p. 12) which sends employees a message that "you can't be trusted, we don't respect your judgement, you aren't very smart, you are incapable of doing this yourself" (Maslach and Leiter, 1997, p. 12).

One Statistics Canada survey found that older workers were four times more likely to retire early if they felt they had little or no job autonomy (Turcotte and Schellenberg, 2005). As a result, organizations intent on keeping valuable older employees must consider giving them more leeway to make decisions.

When workload is a problem, allowing employees some latitude in choosing how they will schedule additional hours, for example, is one way to provide more control. Another is to allow employees to have some say in what types of additional work they will take on. Workers who are given more control over their environment and the way in which they carry out the work are also more likely to report greater satisfaction with the job and to be able to mitigate the stress of demanding jobs.

Job domain 3: reward

The third of Leiter and Maslach's (2005) six job domains – reward – is also a thorny issue for management. Organizations have typically used money as a reward for performance either in the way of salary or bonuses. Nonetheless, reward programs are approached, like issues of control above, in a "narrow, cookie-cutter, one-size fits all-approach" (Maslach and Leiter, 1997, p. 12).

They are narrow and mechanical in that they overlook equally important ways to reward employees. For example, in an age of lay-offs and downsizing, job security is increasingly seen as a material reward. For some employees, career advancement, greater responsibility (accompanied by the requisite support) and/or control over how the job is done are also seen as rewards. In other cases, a reward might involve adjusting an employee's schedule in order to accommodate continuing education or training. Some kind of brief downtime (i.e., time away from the most hectic part of the job), or time away from the job itself may also be desirable.

Research over the years has shown that employees of all ages value non-monetary types of reward as much as money. Hewlett et al. (2009), for example, found similarities among the most junior and most senior employees regarding the rewards considered as important as money. "Gen Y" workers (those born between 1980 and 1994) were looking for such rewards in the following order:

- High-quality colleagues
- Flexible work arrangements
- Prospects for advancement
- Recognition from one's company or boss
- A steady rate of advancement and promotion
- Access to new experiences and challenges. (Hewlett et al., 2009, p. 76)

Along similar lines, "Baby Boomers" (those born between 1945 and 1960) were looking for these rewards in the following order:

- High-quality colleagues
- An intellectually stimulating workplace
- Autonomy regarding work tasks
- Flexible work arrangements
- Access to new experiences and challenges
- Giving back to the world through work
- Recognition from one's company or boss. (Hewlett et al., 2009, p. 76)

The good news for organizations, then, is that it's not all about the money. Satisfying an employee's compensation needs is, indeed, just one way of keeping an employee happy. For there to be a 'fit' in the organization with regards to reward, however, organizations must also keep in mind other employee aspirations which are often as important as salary itself.

Job domain 4: community

Number four of Leiter and Maslach's (2005) six job domains – community – becomes a problem for employers when employees are unable to find positive social contacts and shared values. Research has shown that employees who have high-quality relationships or positive connections with co-workers are more likely to value and enjoy the workplace, and less likely to burn out over time (Fernet et al., 2010). High-quality relationships are defined as those which are "enriching, harmonious, satisfying, and that inspire trust" (Fernet et al., 2010; Senécal et al., 1992). Relationships of this sort act as a buffer for stress; in contrast, conflict with co-workers or feelings of isolation on the job are more likely to result in burnout. The increasing trend towards contract and short-term workers in workplaces creates a problem for building and sustaining community since full-timers see little point in cultivating relationships with people who may only be working temporarily.

A study by career consultancy firm Sanders and Sidney reported that the majority of a worker's friends are now made on the job. Friends are also a huge source of job satisfaction; like pay and benefits, they help to keep employees from moving to other jobs. According to the study, 80 percent of employers realize they should be working to create a positive atmosphere at work; however, most don't understand that doing so is better for business (Hilpern, 2000; Collins, 2000) and good for the bottom line.

For organizations to benefit from cohesive employee relationships, they need to understand something about what it is that promotes or undermines community amongst employees. Some workplaces, for example,

> seem designed to promote conflict between co-workers rather than cooperation. Co-workers may find themselves in competition with each other for favored positions, recognition, bonuses, and promotions. If this is the case, one-upmanship, backbiting, and putdowns are the order of the day. People are concerned about "me first" and may try to look good by making their peers look bad. Obviously, there are few attempts to help "the other guys" since there are no rewards for doing so and the helper only gets left behind as they move up. Furthermore, people are unwilling to ask for help or share their feelings, since doing so is often interpreted as a sign of incompetence or weakness. It may be used against them later on when a promotion report notes a lack of independence and overemotionality . . . The lack of trust that exists in such job settings puts invisible walls between potential allies. (Maslach, 2003, p. 70)

Employers need to do what they can to bring down these invisible walls. To begin, they can encourage a sense of community by sponsoring more social activities, allowing for a relaxed dress code and designing work and social areas in such a way that they encourage positive interaction.

Employees must play a part in designing these social activities so that it isn't seen as something imposed top-down which they must endure to keep management happy. By allowing employees to set aside a small amount of time each week to come together and do some activity that interests them, employers increase the possibility of making productive friendships in the workplace, and productive friendships lead to a better atmosphere, greater retention of valuable employees, and increased motivation.

Nevertheless, building a better community in the workplace involves more than just workplace friends, it also requires supportive managers. Research has shown that in workplaces lacking supportive managers, employees are more likely to have higher emotional exhaustion (Edwards et al., 2001; Hannigan et al., 2000) and more likely to have negative attitudes towards the individuals whom they serve (Hannigan et al., 2000). Where managers are given training in providing additional support to their subordinates, however, those subordinates not only had greater job satisfaction but also fared better both physically and psychologically (Deci et al., 1989). Employees who are treated with respect and made to feel valued in the workplace have less difficulty handling heavy work demands than their less respected co-workers (Galt, 2006). They are also less likely to want to quit; researchers have discovered that when employees see their supervisors as supportive, they are less likely to pack in the job (Leung and Lee, 2006).

Job domain 5: fairness

Fairness, the fifth of Leiter and Maslach's (2005) six job domains, is a broad concept which covers various types of inequity in the workplace including the amount of work, remuneration for work, performance evaluations, and promotions. Fairness issues also include favoritism, discrimination, lying, cheating, and the way in which complaints or grievances are dealt with by the organization.

Fairness is hardwired into our systems, and treatment which is unfair touches a very old and primal part of our being. Research has shown that the amygdala, an ancient primitive part of the brain which controls feelings of anger and fear, also reacts strongly and automatically to issues of fairness. Specifically, researchers have found that humans react both instantly and aggressively when others are perceived to be acting unfairly. Individuals who perceive themselves as being treated unfairly will respond by making decisions that hurt the person acting unfairly even if it also means hurting themselves. For example, in studies involving how to split a certain sum of money, where the person in charge takes 80 percent for himself and leaves 20 percent for the other individual, that other individual will often refuse to take his 20 percent share even though it is not in his best interests to come away with nothing (ScienceDaily, 2011).

As a result, employers who act in ways which are perceived as unfair are likely to create a work environment in which employees not only act irrationally but are often openly hostile even if it means they are also hurting themselves by responding this way. To counteract this, for example, promotions should be handled fairly, and employees should have some understanding not only as to why one individual was promoted but why others weren't. The reasons behind decisions regarding distribution of work, pay and other resources should also be communicated openly, and the opportunity provided for frequent and constructive feedback to employees.

Workers who feel they are being passed over for promotion or other rewards should not be ignored. Instead, supervisors should consider having them attend certain courses in skills and professional development or provide them with some other form of mentoring. Procedures should also be put in place to ensure that the same "high flyer" employees are not feted over and over again but that other employees have a chance to be recognized and to shine as well.

In addition to procedures to deal with the meting out of rewards, all organizations should have policies in place to handle harassment, bullying and other forms of abusive behavior which can also lead to inequitable treatment. Abusive behavior, discrimination and favoritism can have implications for employee behavior and morale that can reverberate throughout the organization and beyond.

These behaviors on the part of management can also lead to trouble with the law. Most jurisdictions in the industrialized world prohibit abusive behavior or bullying on the job. These laws also prohibit discriminatory behavior and various government agencies, boards and judicial and quasi-judicial bodies exist to deal with complaints involving discrimination. Organizations which are unable to handle these problems internally sometimes find themselves exposed publically and dealing with the negative publicity that accompanies it. Through education and training, management must understand what kind of behaviour is unacceptable in the workplace and employees must be provided with information regarding what they can do if they are the target of this kind of behavior. Similarly, both employees and supervisors must be given training in cultural diversity in order to avoid unpleasant incidents which lead to trouble with the law.

Having clear procedures in place in an organization is crucial as is open communication. Some employers, for example, have also experimented with weekly group sessions in which employees were encouraged to discuss any perceived inequities in the workplace (van Dierendonck et al., 1998). When these employees were later compared with a control group, the weekly group participants reported a significant decrease in emotional exhaustion which could lead to burnout, and an increase in perceived equity within the organization.

Job domain 6: values

Finally, the sixth of Leiter and Maslach's (2005) six job domains – values – is also problematic even though at first glance this would not appear to be the case. Indeed, organizations like listing their values in their public documents and/or on their websites; doing so allows them to post heartwarming and inspirational statements about their goals and vision. Real values, however, have little to do with heartwarming words on promotional materials, and more about beliefs and opinions on what matters in the workplace. In order to know what an organization really values, one has only to look at the behavior which it rewards and punishes. This translates into a problem with fit when employees experience conflict between their individual or professional values and those of the organization, or between what is expected of them in the job and what they feel is the right thing to do.

For example, organizations which claim to value integrity and yet reward dishonesty, or which claim to value input and open communication and then develop plans in secret are sending a specific message to employees. Similarly, organizations which assert they care about work–life balance and then only reward employees who work long hours, or those which claim to have the client foremost in mind and then make decisions which hinder the delivery of frontline service are letting employees know what they really value. Management sows the seeds for discontent and burnout when it claims a commitment to quality, service and excellence, then undermines it in these and other ways.

Some experts suggest that organizations may occasionally want to conduct a "values inventory" or assessment in order to uncover any discrepancies between perceived and actual values (Benedictine University, 2012). Understandably, management and employees are likely to view a values inventory or survey as unimportant or as some "hoop jumping" exercise. Nonetheless, both organizations and employees can benefit not only when values are re-assessed honestly from time to time but also when they are reinforced. So, for example, if work–life balance is truly important (and not just something to which the organization pays lip service), then it is vital that supervisors talk to employees who are in the office day and night about going home at a decent hour. When management and employees take the time to define values and then to reinforce them with consistent behavior, employees are less likely to experience the stress that mixed messages send and less likely to have it affect their work performance. Also, organizations are less likely to have to address problematic behavior when values are clearly understood and reinforced in the workplace.

Conclusion

This chapter has explored mismatches in six job domains – workload, control, reward, community, fairness, and values – from the perspective of the worker as well as that of the organization and has suggested ways of overcoming problems from each. Mismatches must be addressed; according to Maslach (2003), not to do so is to expose employees to burnout and to leave organizations more vulnerable to unpleasant after-effects.

Nonetheless, as this chapter has shown, there is much that both individuals and organizations can do to repair the problems which arise in each one of these domains. Individuals can, for example, protect themselves by staying healthy, drawing a clear line between work and home, and responding to stressors on the job in a way that will minimize their effects.

But employees alone can't do it all. Organizations must understand that the sick or dysfunctional work environment affects not only the worker, but also the work and the bottom line. By showing leadership and taking measures to bring about or restore a healthy work environment, organizations ensure their own success as well.

References

25 percent bored in office job (12 January 2012) *METRO*, 6.

30 percent of workers' values mismatch employers' (20 April 2007) *The Globe and Mail*, C2.

Aiken, L.H., Clarke, S.P., Sloane, D.M., Sochalski, J., and Silber, J.H. (2002) Hospital nurse staffing and patient mortality, nurse burnout, and job dissatisfaction. *Journal of American Medical Association*, 288(16), 1987–93.

American Running Association (1 July 2011) Exercise and stress: Work out to work it out. *Running & FitNews*, 29(4). Retrieved from *http://static.americanrunning.org/fitnews/ARAfitnews_V29_4/index.html.*

Aon Consulting (8 February 2010) Rewards and their influence in the workplace. Retrieved from *http://www.ecplaza.net/news/rewards-and-their-influence-in-36276-14.html.*

Awa, W., Plaumann, M., and Walter, U. (2010) Burnout prevention: A review of intervention programs. *Patient Education and Counselling*, 78(2), 184–90.

Baird, T., and Baird, Z. (2005) *Running on empty: Dealing with burnout in the library setting.* LISCareer.com. Retrieved from *http://www.liscareer.com/baird_burnout.htm.*

Beauchesne, E. (30 August 2006) Meaningful work "better than money". *National Post*, WK4.

Benedictine University, Center for Values-Driven Leadership (2012) Organizational values inventory – A look at where leadership is heading. Retrieved from *http://valuesdrivenleadership.blogspot.ca/2011/05/organizational-values-inventory.html#more.*

Brody, J. (1 June 2011) Importance of a good night's sleep; As people grow older, a host of major benefits hinges on getting rest. *International Herald Tribune*, 8.
Bruzzese, A. (9 June 2010) On the job: Meaningful work helps keep you happy after you've gone home. *USA Today*. Retrieved from *http://usatoday30.usatoday.com/money/jobcenter/workplace/bruzzese/2010-06-09-meaningful-work_N.htm*.
Casserley, T.B., and Megginson, D. (2009) *Learning from Burnout: Developing Sustainable Leaders and Avoiding Career Derailment*. Oxford, UK: Butterworth-Heinemann.
Censky, A. (18 May 2012) Vacation? No Thanks, Boss. CNNMoney.com. Retrieved from *http://money.cnn.com/2012/05/18/news/economy/unused_vacation_days/index.htm*.
CNNMoney.com (2012) 100 best companies to work for 2012: Best benefits: Sabbaticals. Retrieved from *http://money.cnn.com/magazines/fortune/best-companies/2012/benefits/sabbaticals.html*.
Collins, V. (12 October 2000) Workplace friendships are good for us – Report urges employers to encourage a sense of community. *The Herald*, GH 13.
Day, A.L., Sibley, A., Scott, N., Tallon, J.M., and Ackroyd-Stolarz, S. (2009) Workplace risks and stressors as predictors of burnout: The moderating impact of job control and team efficacy. *Canadian Journal of Administrative Sciences*, 26, 7–22.
Deci, E., Connell, J., and Ryan, R. (1989) Self-determination in a work organization. *Journal of Applied Psychology*, 74, 580–90.
Dixon, R.B., and Wills, R.L. (1 February 2010) Take your vacation. Please. When you take time off, everyone wins. *Corporate Counsel*, 17(2), 43.
Edwards, D., Burnard, P., Coyle, D., Fothergill, A., and Hannigan, B. (2001) A stepwise multivariate analysis of factors that contribute to stress for mental health nurses working in the community. *Journal of Advanced Nursing*, 36(6), 805–13.
Ekstedt, M. (2005) *Burnout and sleep* (Thesis). Department of Public Health Sciences, Karolinska Institute, Stockholm, Sweden.
Elson, P. (16 April 1979) A break in routine helps beat boredom. *The Globe and Mail*, S23.
Etzion, D. (2003) Annual vacation: Duration of relief from job stressors and burnout. *Anxiety, Stress and Coping*, 16(2), 213–26.
Expedia.ca (30 November 2011) Canadians can't let go of work even when on vacation. Retrieved from *http://www.expedia.ca/daily/enc4105/service/press/releases/2011/1111-Vacation-Deprivation.asp*.
Expedia.co.uk (2012) All work no play. Retrieved from *http://inside.expedia.co.uk/travel-trends/vacation-deprivation-57*.
Expedia.com (2011) *2011 vacation deprivation study*. Expedia.com. Retrieved from *http://media.expedia.com/media/content/expus/graphics/other/pdf/vacation-deprivation-fact-sheetnov2011.pdf*.
Fernet, C., Gagne, M., and Austin, S. (2010) When does quality of relationships with coworkers predict burnout over time? The moderating role of work motivation. *Journal of Organizational Behavior*, 31, 1163–80.
Fujiwara, K., Tsukishima, E., Tsutsumi, A., Kawakami, N., and Kishi, R. (2003) Interpersonal conflict, social support, and burnout among home care workers in Japan. *Journal of Occupational Health*, 45(5), 313–20.

Galt, V. (12 August 2006) New study sheds light on preventing burnout. *The Globe and Mail*, B10.

Gardner, M. (24 September 2007) Bosses see benefit in giving sabbaticals to employees. *Christian Science Monitor*, 13–16.

Giancola, F. (1 July 2006) Making sense of sabbaticals. *Workspan*, 39–41.

Glicken, M.D., and Janka, K. (1982) Executives under fire: The burnout syndrome. *California Management Review*, 24(3), 67–72.

Glouberman, D. (2002) *The Joy of Burnout.* Isle of Wight: Skyros Books.

Goleman, D. (13 May 1986) Relaxation: Surprising benefits detected. *New York Times.* Retrieved from *http://www.nytimes.com/1986/05/13/science/relaxation-surprising-benefits-detected.html?pagewanted=all.*

Greenglass, E.R., Burke, R.J., and Konarski, R. (1997) The impact of social support on the development of burnout in teachers: Examination of a model. *Work & Stress*, 11, 267–78.

Gupta, S., Paterson, M.L., Lysaght, R.M., and von Zweck, C.M. (2012) Experiences of burnout and coping strategies utilized by occupational therapists. *The Canadian Journal of Occupational Therapy*, 79(2), 86–95.

Hannigan, B., Edwards, D., Coyle, C., Fothergill, A., and Burnard, P. (2000) Burnout in community mental health nurses: Findings from the All-Wales Stress Study. *Journal of Psychiatric and Mental Health Nursing*, 7, 127–34.

Hewlett, S.A., and Luce, C.B. (2006) Extreme jobs: The dangerous allure of the 70-hour workweek. *Harvard Business Review*, 84(12), 49–59.

Hewlett, S.A., Sherbin, L., and Sumberg, K. (2009) How Gen Y and Boomers will reshape your agenda (Special Issue: Managing in the New World). *Harvard Business Review*, 87(7–8), 71–6.

Hilpern, K. (29 October 2000) The friendship factory: Work mates are our best mates, and firms will prosper if they create a sense of community. *The Independent.* Retrieved from *http://www.independent.co.uk/news/business/news/the-friendship-factory-635581.html.*

Hoare, R. (2 May 2012) Is workplace boredom the new stress? Retrieved from *http://edition.cnn.com/2012/05/02/business/workplace-boredom-stress/index.html.*

Hultman, K. (2005) Evaluating organizational values. *Organization Development Journal*, 3(4), 32–44.

Hyrkas, K. (2005) Clinical supervision, burnout, and job satisfaction among mental health and psychiatric nurses in Finland. *Issues in Mental Health Nursing*, 26, 531–56.

Immen, W. (29 June 2012) Learn to delegate or risk diluting skills. *The Globe and Mail.* Retrieved from *http://m.theglobeandmail.com/report-on-business/small-business/sb-managing/succession-planning/learn-to-delegate-or-risk-diluting-skills/article4381345/?service=mobile.*

Joyce, A. (8 October 2005) Stressed out by having nothing to do: Busiest employees are the happiest. *The Washington Post*, D12.

Kanji, N., White, A., and Ernst, E. (2006) Autogenic training to reduce anxiety in nursing students: Randomized controlled trail. *Journal of Advanced Nursing*, 53, 729–35.

Karasek, R.A. Jr. (1979) Job demands, job decision latitude, and mental strain: Implications for job redesign. *Administrative Science Quarterly*, 24(2), 285–308.

Kaufman, B. (2005) Beating stress and preventing burnout. *Handbook of Business Strategy*, 6(1), 171–5.

Keeton, K., Fenner, D.E., Johnson, T.R.B., and Hayward, R.A. (2007) Predictors of physician career satisfaction, work–life balance, and burnout. *Obstetrics and Gynecology*, 109, 949–55.

Leiter, M.P., and Maslach, C. (2005) *Banishing Burnout: Six Strategies for Improving Your Relationship with Work*. San Francisco, California: Jossey-Bass.

Leung, D.Y.P., and Lee, W.S. (2006) Predicting intention to quit among Chinese teachers: Differential predictability of the components of burnout. *Anxiety, Stress & Coping*, 19(2), 129–41.

Maslach, C. (2003) *Burnout: The Cost of Caring*. Englewood Cliffs, New Jersey: Prentice-Hall. (Original work published 1982.)

Maslach, C. and Leiter, M.P. (1997) *The Truth about Burnout: How Organizations Cause Personal Stress and What to Do about it*. San Francisco, California: Jossey-Bass.

Maslach, C., Schaufeli, W.B., and Leiter, M.P. (2001) Job burnout. *Annual Review of Psychology*, 52, 397–422.

Moon, M.A. (2009) Mindfulness program eases physician burnout. *Clinical Psychiatry News*, 7(10), 57.

Much worker burnout traced to poor supervision (14 December 1992) *The Toronto Star*, B1

Naish, J. (24 November 2005) Killer bosses. *The Times*, Career 4.

Nevanpera, N.J., Hopsu, L., Kuosma, E., Ukkola, O., Uitti, J., and Laitinen, J.H. (2012) Occupational burnout, eating behavior, and weight among working women. *The American Journal of Clinical Nutrition*, 95(4), 934.

Pines, A. and Aronson, E. (1988) *Career Burnout: Causes and Cures*. New York, New York: The Free Press.

Ray, A. (1 February 2012) 5 ways employers can reduce worker stress. PropertyCasualty360. Retrieved from *http://www.propertycasualty360.com/2012/02/01/5-ways-employers-can-reduce-worker-stress*.

Rowe, M. (2000) Skills training in the long-term management of stress and occupational burnout. *Current Psychology*, 19(3), 215–28.

Sala, H.J. (2009) *Making Your Emotions Work for You: Coping with Stress, Avoiding Burnout, Overcoming Fear . . . and More*. Eugene, Oregon: Harvest House Publishers.

Salmela-Aro, K., Nataanen, P., and Nurmi, J.E. (2004) The role of work-related personal projects during two burnout interventions: A longitudinal study. *Work & Stress*, 18, 208–30.

Schaufeli, W.B. and Enzmann, D. (1998) *The Burnout Companion to Study and Practice: A Critical Analysis*. London, UK: Taylor and Francis.

ScienceDaily (3 May 2011) Sense of justice built into the brain, imaging study shows. *ScienceDaily*. Retrieved from *http://www.sciencedaily.com/releases/2011/05/110503171743.htm*.

Senécal, C.B., Vallerand, R.J., and Vallieres, E.F. (1992) Construction et validation de l'Echelle de la qualité des Relations Interpersonnelles (EQRI) [Construction and validation of the Quality of Interpersonal Relationships Scale (QIRS)]. *Revue Européenne de Psychologie Appliquée*, 42, 315–22.

Senior, J. (26 November 2006) Can't get no satisfaction: In a culture where work can be a religion, burnout is its crisis of faith. *New York Magazine*. Retrieved from *http://nymag.com/news/features/24757/*.

Smith, R. (5 May 2010) People who sleep for less than six hours "die early". *Telegraph*. Retrieved from *http://www.telegraph.co.uk/health/healthnews/7677812/People-who-sleep-for-less-than-six-hours-die-early.html*.

Söderström, M., Jeding, K., Ekstedt, M., Perski, A., and Åkerstedt, T. (2012) Insufficient sleep predicts clinical burnout. *Journal of Occupational Health Psychology*, 17(2), 175–83.

Spector, P.E. (1986) Perceived control by employees: A meta-analysis of studies concerning autonomy and participation at work. *Human Relations*, 39, 1005–16.

Steckler, J. (28 May 2012) Manage stress with proper diet. *Chicago Daily Herald*, 2.

Talbott, S. (2012) *Secret of Vigor: How to Overcome Burnout, Restore Metabolic Balance, and Reclaim Your Natural Energy*. Alameda, California: Hunter House.

Tepper, B. (2000) Consequences of abusive supervision. *Academy of Management Journal*, 43, 178–90.

Toker, S., and Biron, M. (2012) Job burnout and depression: Unraveling their temporal relationship and considering the role of physical activity. *Journal of Applied Psychology*, 97(3), 699–710.

Tunick, J. (2000) Preventing burnout. *Civil Engineering*, 70(12), 66–7.

Turcotte, M., and Schellenberg, G. (2005) Job strain and retirement. *Perspectives on Labour and Income*', (Statistics Canada – Catalogue no. 75-001-XIE), 6(7), 13–17.

UCIrvine Today (3 May 2012) Email "vacations" decrease stress, increase concentration. Retrieved from *http://today.uci.edu/news/2012/05/nr_email_120503.php*.

Vaccaro, P.J. (1998) Balancing act: Five ways to say "no" effectively. *Family Practice Management*, 5(7), 71–2.

van Dierendonck, D., Schaufeli, W.B., and Buunk, B.P. (1998) The evaluation of an individual burnout intervention program: The role of inequity and social support. *Journal of Applied Psychology*, 83(3), 392–407.

Van Rhenen, W., Blonk, R.W.B., Van der Klink, J.J.L., Van Dijk, F.J.H., and Schaufeli, W.B. (2005) The effect of a cognitive and a physical stress-reducing programme on psychological complaints. *International Archives of Occupational and Environmetal Health*, 78, 139–48.

Verissimo, R., Sa, L.O., and Barreto, J. (2003) Control as an approach to preventing burnout among health care personnel. *Journal of Psychosomatic Research*, 55, 147–78.

Ward, M. and Cowman, S. (2007) Job satisfaction in psychiatric nursing. *Journal of Psychiatric and Mental Health Nursing*, 14, 454–61.

Wiener, L. (1 August 2007) Yes, you're stressed . . . but what are you doing about it? *Shape*. Retrieved from *http://www.redorbit.com/news/health/1125962/yes_youre_stressed_but_what_are_you_doing_about_it/*.

Wilkins, K. and Beaudet, M.P. (1998) Work stress and health. *Health Reports* (Statistics Canada, catalogue no. 82-003-XPB), 10(3), 47–62.

Work boredom "may be responsible for employees calling in sick" (21 February 2012) *The Indian Express*. Retrieved from *http://m.indianexpress.com/news/work-boredom-may-be-responsible-for-employees-calling-in-sick/914832/*.

Yagil, D. (2006) The relationship of abusive and supportive workplace supervision to employee burnout and upward influence tactics. *Journal of Emotional Abuse*, 6(1), 49–65.

9

Managing and overcoming burnout

Abstract: This chapter discusses how to manage and overcome burnout. It explores the initial steps which should be taken by an employee including seeking help in the form of social support, a trained professional, an employee assistance program, and career counseling. It also discusses the importance of lowering stress, taking the time to recover, understanding and accepting the feelings and expectations that accompany burnout, and mending fences. Finally, the importance of restructuring the current job is discussed, along with the possibility of finding another job when all else has failed.

Key words: burnout, managing burnout, overcoming burnout.

Introduction

At some point, the question may no longer be how to avoid burnout, but how to cope once it has already established itself. Discovering that you are burned out, experts tell us, serves as

> a wakeup call that your life and career are unsustainable. The fact of burning out tells you that your current way of operating at work and in your career has become untenable. It indicates that the story you have constructed about who you are, your purpose in being here and what gives you meaning, is no longer appropriate for the context in which you find yourself. (Casserley and Megginson, 2009, p. 109)

Burnout doesn't have to be viewed entirely negatively, though. Psychotherapist Dina Glouberman has written a book on "the joy of burnout," and views it as an important potential turning point in our lives. She writes,

> [b]urning out could almost be defined as joylessness. We have lost contact with the Divine Comedy, that ability to laugh compassionately at the terrible and wonderful ways in which life works itself out. Yet burnout forces us to take a step towards joy. We won't stop, so burnout stops us. We won't make a space for ourselves, so we burn out and all we have is space. And it is out of the space that the joy eventually comes. (2002, p. 11)

For Glouberman, burnout is a hard lesson that most people ultimately find themselves grateful to have learned. It stops us from continuing to live in an unsustainable way; it also makes us look back and wonder how we could have ever managed to get away from that unsustainable life if burnout had not stopped us in our tracks.

So, for Glouberman and others, the good news is that even though it might take time, workers can come back from even the worst cases of burnout by making the necessary adjustments in their lives. This means changing priorities, habits, and ways of thinking about work. It also involves finding new ways of coping or adapting existing coping methods to the circumstances. In some cases, it might mean changing jobs altogether.

Whatever the steps taken to minimize or eliminate the sources of burnout, it is quite possible to recover and come away from the syndrome without experiencing significant detrimental long-term consequences (Cherniss, 1992). Individuals who have burned out can come back wiser and stronger than they were before. Recovery generally doesn't happen overnight, however; researchers have determined that one to three years are often necessary to recover from burnout (Bernier, 1998).

In a workplace in which the employer has employee assistance available (i.e., through management, human resources, wellness or employee assistance programs (EAP)), the burned-out worker doesn't have to go it alone. Workplace assistance personnel and programs will help to match the employee in distress with specific sources of help such as internal counseling programs, counselors external to the organization or other relevant assistance programs.

Some workplaces, however, might be too small to have large and sophisticated human resources departments and EAP programs, while others might not be receptive to helping employees with stress and burnout. In these cases, employees may have to make things better for themselves without the benefit of workplace assistance. Luckily, experts offer numerous suggestions on how to recover with or without assistance, and these are discussed below.

Development, derailment or a new path to joy?

In their research, Casserley and Megginson (2009) found that burnout affected approximately 20 percent of the individuals who were part of the high potential talent pool at four different organizations. This was the case regardless of the nationalities of the individuals interviewed or the country in which they worked. Burnout in most of these led to "career derailment" (Casserley and Megginson, 2009) or some other form of career setback – individuals left their jobs, could no longer keep up because of illness, or were denied a promotion.

Casserley and Megginson observed that the difference between burnout serving as a "significant developmental experience" (i.e., learning and benefitting in some other way from the experience of burnout) and "derailing" had to do with whether the burned out individuals could accept some responsibility for their condition. As one interviewee noted:

> *I felt I had too much work, a rubbish boss and no support – I felt no one noticed or cared. I felt like a victim. I couldn't see my way out. So whilst I recognized that I had to change something – I had endless conversations with my husband about how to change – I didn't feel empowered to do it. I knew I had to change the way I was working, but the very recognition of that felt like more pressure, one more thing to deal with. It felt terribly hard. But as I eventually started resolving it, what started to give me hope was my beginning to realize I wasn't a victim, but that I actually had some choice and some responsibility in the situation.* (Casserley and Megginson, 2009, p. 96; italic in original)

Accepting responsibility in this case has to do with recognizing that one's ability to carry on is not limitless, and that one must participate in trying to fix the problem.

For Glouberman (2002), burnout, whether it ends a career or results in a significant learning experience, is the beginning of a brand new life and new possibilities. Burnout, she notes, is

> stopping doing what is familiar and has gained us rewards in the past. We live in a culture in which stopping often signifies a failure to make the grade, to continue up the ladder, to win the prizes that are hanging at the top, or at least on the rung above us. As long as we are climbing, we are holding on to our old identities or old beliefs about what we should be, and winning for ourselves the right to have new and more sparkling identities. If we lose these, who and what will we be? Yet once we get to burnout, we often have no choice but to stop. Stop is of course what we need to do most of all. We need to stop fighting the burnout, indeed stop any kind of struggle. It is time for a new beginning. (p. 150)

Burnout "seems to be a state of joylessness" but is actually "pointing the way to joy" (p. 153). It forces us to stop doing something that was sucking the life out of us and to find a new way to live our lives.

As noted above, organizations differ in terms of their willingness to help employees suffering from burnout as well as in terms of the resources which they have at their disposal. Some organizations have sympathetic and resourceful human resources departments or assistance programs which are able to offer advice, provide counseling and put into place practical measures which are designed to combat burnout and eliminate, diminish or provide coping mechanisms for the sources of stress responsible for the condition.

Other organizations may not be as helpful. Depending on the workplace culture, there might be very little patience for employees who cannot keep pace, no matter how frantic or hectic. For workers in this type of environment, burnout results in a loss of confidence and in feelings of being trapped. Work suffers and employees see themselves merely going through the motions, afraid to admit to burnout, and worried about seeking help or counseling. Not surprisingly, letting management know that there is a problem sometimes results in no longer being considered for promotion, and even being removed from the high-potential talent pool.

With or without help from an employer, the first step on the path to recovery is to determine the causes of burnout and, if needed, to get some help. Sometimes separating the toxic from the garden-variety forms of stress on the job can be difficult, and this is where a trained and trusted individual can help. In addition to pinpointing the problem, a professional can provide advice on re-establishing balance. How to start down the path towards healing, and how to take those first steps are discussed below.

First steps

a) Determine the cause

As noted above, where it's not already obvious, employees who are burning out need to identify the specific sources of stress. When the workplace has numerous sources of stress, though, it can be hard to work out which are manageable and which put employees at risk for burnout.

One way to sort this out is by keeping a stress log or diary. Too much work, for example, might be a source of stress one day, while conflict with a colleague or superior might cause stress the next. The question is whether a pattern emerges as to overwhelming sources of stress, and to determine this, it's best to keep track of the frequency or severity of the stressful events and the conditions under which they occur.

Stress diaries can take a number of different forms. Bruce (2007), for example, suggests using a notebook and making seven columns. The columns should be titled "date, time, situation, scale, symptoms, efficiency and reaction" (Bruce, 2007). She then recommends making notes in the diary several times a day if possible.

The columns for "date" and "time" are self-explanatory. The "situation" column requires the note-taker to write down the stressful occurrence. Bruce says that the employee must be honest about what is stressful. The point is to observe the situation as objectively as possible and then to simply record it.

For "scale," Bruce suggests giving the stressful event or occurrence a score between one and ten with one being the least stressful and ten being the most stressful. The response of the diary writer to the stressful event should also be recorded in the "symptoms" column. For example, if the situation resulted in anger, tears, frustration, etc., all of this should be jotted down.

The "efficiency" column comes next. This column requires the diary keeper to record honestly and accurately how effectively she is able to focus on the job given the source of stress. A number between one and ten is required here too, with the number one indicating that the individual cannot focus on the job given the source of stress, and with the number ten indicating that focus and work are not a problem and that the job is being done as efficiently as ever.

Finally, the "reaction" of the diary keeper should be recorded. What was the response to the source of stress? Was the problem solved? Is it ongoing? Has it been made worse by doing nothing or, perhaps, by doing something that inflamed the situation?

Bruce suggests that the diary is one way to provide an answer regarding the problem(s) in the workplace. Once the sources of stress are identified, it is easier to have some idea as to whether they are solvable and in what way. Researchers have found that keeping a diary not only aids in identifying the problem, but also allows for self-reflection. It has the additional benefit of being cathartic and increasing job satisfaction (Clarkson and Hodgkinson, 2007).

b) Talk to others

Individuals who are burning out generally don't recognize the symptoms; only in hindsight is there some recognition of what was going on at the time. Instead, caring friends, family and/or co-workers tend to see what is happening more clearly to their friend or loved one even if they are at a loss to understand or resolve the problem. These caring individuals are often able to notice changes in behavior, feelings, thoughts and health that signal a real problem. They also act as a sounding board, allowing the employee in distress to try to pinpoint the problem and to work out how to resolve it.

Of course, making use of this support can be difficult when an employee is burning out. For example, social support from co-workers can be difficult to find in an organization in which there is, at best, little cohesiveness and, at worst, great animosity among workers. More to the point, even where there might be cohesiveness and support, individuals who are burning out are more inclined to withdraw rather than seek help from their supervisors or colleagues. Even those people who should be most important in their lives – family and friends – can seem like yet another burden rather than a source of relief and support.

Nonetheless, someone who has similar values and priorities and who can provide emotional support and/or listen actively without being critical or judgmental is what the at-risk employee needs (Pines and Aronson, 1988). That person can be a family member, friend, physician, therapist, naturopath, co-worker, colleague, supervisor or other contact. This type of support is discussed in more detail below.

c) Social support

Social support is critical to surviving and coming back from burnout. Casserley and Megginson (2009) note that individuals who ultimately view burnout as a "significant developmental experience" rather than a "derailment," have had a relationship with someone throughout the experience – a trusted colleague, a significant other, a counselor or someone else – who has listened to them and sustained them during periods of overwork, long hours and high stress. At the same time, that person may challenge the individual in distress and hold him or her to account for and accept some responsibility for the dilemma.

Glouberman (2002) recommends finding a "soul community" made up of people who have no vested interest in "who we used to be" (p. 222); in other words, these people have no vested interest in who we were *vis-à-vis* our jobs. They give us space and time; they care for us and are respectful about who we are becoming as we move beyond burnout. At the same time, as we give up the old habits that burned us out in the first place, they do not harm us by introducing us to destructive new habits involving alcohol, drugs, etc. These soul communities can come about through therapy or by joining a support group. They also may be made up of family and friends.

Over and above community, family and friends, social support in the workplace is often key to avoiding or surviving burnout. Research is clear, first of all, that effective workplace social networks reduce the incidence of burnout. When relations with co-workers are good,

> professionals experiencing stress often [turn] to others for advice, comfort, tension reduction, help in achieving distance from the situation or in intellectualizing it, and a sense of shared responsibility . . . burnout [is] less severe in those institutions that [allow] staff to express their feelings, get feedback and support from others, and develop new goals for themselves and their clients than in those institutions that [do] not allow it. (Pines and Aronson, 1988, p. 171)

In institutions where workers enjoy each other's company, get along, and support each other, workers report that they have many more good days than bad (Pines and Aronson, 1988). This is important in keeping burnout at bay; it is also crucial to coming back from burnout.

Where social support from community, family, friends or colleagues is not available, or where that in and of itself is not enough to recover, a burned out employee might also want to find professional help in the way of a physician, therapist or other like professional.

d) Find a therapist

A professional therapist will help identify the problem and then help the employee to find some way to react logically to the organizational conditions responsible for burnout (Glicken, 1983). Workers need to know what they expect of their work, the meaning and importance they attach to it, how they plan to achieve their expectations, the amount of work necessary to achieve what they want, and the amount of work and stress they feel they are able to handle. Glicken (1983) notes that employees in distress are often "locked in place emotionally" and lack confidence. They mistrust their own judgment to such an extent that their therapist must typically provide step-by-step advice.

How the therapist assists can manifest itself in a number of different ways. Clinical psychologist Herbert J. Freudenberger, for example (who was one of the first experts to use the term "burnout" in connection with the phenomenon), believes the first step in the process is having the employee (who is now a patient) recognize the problem. From there, he has the patient focus on exercise, nutrition and diet, and reordering values. Freudenberger tells patients that

> they must change their priorities. . . . Instead of putting all their time and energy into work, they must stop and shift their concerns to more personal matters – family, friends, hobbies, whatever. Anything that will take their mind away from the workplace and closer to home. ("Psychological counselling," 1986)

In contrast, psychiatrist Thomas Kalman believes that expectations are the primary problem and tailors his sessions to dealing with the patient's unrealistic self-image. "I try to get patients to recognize that their perfectionist images and expectations are unreal," he explains. "And once they can accept their own limitations, they are well on the way to cure" ("Psychological counselling," 1986).

Contrasted with these approaches, some therapists use relaxation, breathing, mindfulness and other individual coping techniques along with active approaches such as communicating with a supervisor in order to enlist help to resolve the problem (Morse et al., 2012). Often a combination of methods, along with more exercise and a better diet, can produce positive results. Nonetheless, as psychologist Stephen Josephson notes, "the balance varies with each individual, and there are no quick or easy cures" ("Psychological counselling," 1986).

Luckily, the treatment offered by mental health therapists and physicians, where properly diagnosed, rarely involves the prescription of medication. Researchers note that individuals who recover from burnout are largely able to do that without requiring anxiolytic or antidepressant drugs (Bernier, 1998). Instead, talking, support and employing the various coping methods discussed are often enough to bring employees back from burnout.

e) Employee assistance and other programs

An employee assistance program (EAP) is intended to be a "no blame," confidential resource for employees who have difficulties which may affect mental or physical health and, consequently, job performance. These programs are relatively common in larger organizations and are designed to provide assistance to help employees solve or cope with their problems, and restore effectiveness on the job. Employees may self-refer, trade unions may refer for their members, or managers can refer the employee when job performance becomes a specific issue. Researchers have discovered that access to an EAP or other form of organizational mental health support is often key in helping employees recover (Bernier, 1998).

Approximately 54 percent of public sector employees and 28 percent of private sector employees in the United States have access to wellness and EAPs (Stoltzfus, 2009). In Canada, roughly two-thirds of worksites have an EAP program (Macdonald et al., 2006).

EAPs offer programs and information intended to help workers at little or no cost to deal with various problems. EAPs can provide written information, audio-visual materials, and access to professionals who are trained to deal with specific types of work-related and mental health problems. They can also offer counseling services.

Assistance is not only aimed at those at risk of burnout; instead EAPs can provide advice to managers who are dealing with an employee who is actually burned out. Where an employee needs time away from work, EAP guidance can provide information to the organization on how to keep in touch with the employee and how to prepare workers, managers and the employee herself for the return to work (Taylor, 2008).

f) Career counseling

Career counseling has been shown to have a significant impact on improving burnout. Studies have revealed that employees who take part in career counseling programs consisting of individual and group counseling sessions show significant improvement in two out of three dimensions of the MBI scale: exhaustion and personal accomplishment (Gorter et al., 2001).

Getting advice from a career counselor might shed some light on aspects of the job that do not fit with the goals, aspirations or personality of the employee. As noted above, participation in a career counseling program has been shown empirically to diminish burnout scores. Career aptitude tests can help determine whether any changes can be made to the current job which would help eliminate sources of stress or lower the risk for burnout. For example, professionals who are asked what the most meaningful activity is in their job, but report that they actually have to spend far more time on activities which they feel are less meaningful, are at greater risk for burnout than those individuals who report spending most of their time on what they feel are their most meaningful activities (Shanafelt et al., 2009). An employee might feel that her most meaningful activity involves working with patrons and that a less meaningful activity involves doing administrative work. In this case, too much time spent doing administrative work is more likely to result in burnout.

Career counseling can also provide advice on whether the job can be tweaked in some way so that it is more suitable to the employee. For example, an employee might be able to approach a supervisor to ask for more of a particular kind of work which the employee finds satisfying. On the other hand, these same tests are sometimes able to determine that an employee is completely temperamentally unsuited to the job and that another line of work would result in a better fit.

g) Lower stress

Once a stress diary and talking to someone has helped pinpoint the source of the stress, the next job is to work on reducing or eliminating that source. As Talbott (2012) notes,

> [c]onventional wisdom and countless commercials bombard us with the idea that the way to get healthy is simply to exercise more and eat a better diet. Both these habits are certainly important parts of being healthy, but from my perspective as a biochemist, I'm going to tell you something you'll hardly hear from anyone else: If you truly want to improve your health, *it is just as important to get your stress levels under control, as it is to eat a healthy diet and get physical activity!* Quite simply, stress has a bigger impact on your life and well-being than almost anything else you encounter. Most people don't understand this fact, or else they ignore it. Worse, some people think they're "tough" enough to handle all the

> stress in their lives. Nothing could be farther from the truth, because stress sets off major biochemical changes in the body. And that is why I call stress the number-one enemy of vigor. (p. 15; italic in original)

Recovering from burnout means shifting perspective and making work–life balance and the diminution of stress a number one priority. Without eliminating the major sources of stress, long-lasting recovery might not be possible (Bernier, 1998).

If the major source of stress is doing too much, then employees need to find a way to do less. If that means no longer being on the fast track to promotion or to the employee-of-the-month list, then that is simply the price that has to be paid for a healthier life. As Sala (2009) notes, "[a]ffirming long-term values comes at the cost of short-term losses" (p. 221).

Glouberman (2002) agrees and notes the importance of correcting the path that we are on. She writes:

> [a]s in the case of social evolution, our underlying personal direction is positive, but working it out can lead us temporarily into conflict, confusion and burnout. The process of evolution is always uneven, with one part of us ahead of another. When our situation changes, it challenges those stuck places and old attachments, and we can temporarily lose our way. If we take a more long-term perspective, we will see that a wrong turning every now and then is part of how we recognize the right path when we get back to it. Burnout is one of those times when we take a wrong turning and stay on it too long. (p. 57)

Looked at in this way, the short-term losses come about as the result of correcting our path as we learn who we really are and struggle to be truer to our real selves.

h) Take the time to recover

Employees who have burned out need time to recover. This means taking active steps to find the time to recover. Where workload is the chief source of stress, for example, it means drawing a clear line between work and home and making time for yourself. Research has revealed that workers who report a higher level of "detachment" from work (i.e., the more a worker is able to feel a sense of distance from work during non-work hours and does not engage in work-related tasks) also report greater life satisfaction in other areas and lower levels of emotional exhaustion than workers who have lower levels of detachment (Fritz et al., 2010). It's not surprising, then, that experts recommend not taking work home, cutting back on hours, and turning off electronic devices after leaving work in order to combat burnout.

One suggestion is to create a routine that marks the beginning of the work day and one that marks the end of the work day. This can be something simple like listening to music on the way in or on the way home (Randstad, 2012) or taking a short walk when you get home. The important thing is to find some routine that clearly signals the beginning and the end of work, and with it, the beginning and the end of the employee's availability in terms of the workplace.

If not working at home is not an option, then another recommendation is to decide on a place and time in your home in which you do not work. This could be an hour at night or some time on the weekend. It could mean that no work is done in the kitchen or dining room when you eat meals. This place and time is meant to rest and recharge by doing something not related to work and by spending time with people who are not part of the work environment.

Time and distance are important in that they allow employees the space and the opportunity necessary to look at what they have experienced and make sense of it. More time away from work also allows the employee to do things important to mental and physical health. This means catching up on sleep, distancing oneself from the source of the stress, working on relaxation techniques, spending more time on personal relationships, and re-learning how to enjoy cultural and social activities along with hobbies and physical exercise (Bernier, 1998). Time alone is also very important (Bernier, 1998).

Where small steps are not enough to recover, however, employees must consider taking more time away. This can take many forms: vacation, sabbatical, sick leave, or exploring the option of part-time work or work-sharing.

For psychotherapist Dina Glouberman, coming back from burnout is all about changing our attitude towards time. This is one of the first and most important steps on what she calls "Radical Healing." Glouberman points out that

> [f]or burnout people, coming back into the present is a matter of life or death. We need all our energy here in order to heal and become

> whole. When we were burning out, we were often running around trying to do the impossible to avert the unavoidable. We were terrified we would run out of time. In fact, out of time is where we need to be. (p. 153)

When we run out of time, she asserts, we finally do what we needed to do all along which is "stop and be silent and in touch with our inner awareness" (p. 155). By doing this, we remember and discover who we are after all this time. We come in touch with our deepest selves, and "recognize that we are at home" (p. 155).

i) Accept your feelings

Experiencing negative feelings such as anger, anxiety, guilt, failure, and disappointment is typical during burnout and they are easier to deal with once employees know why they are feeling this way. It is, in fact, not unusual for individuals to feel, while burning out, that they might be experiencing significant mental health difficulties including madness, paranoia, and depression (Bernier, 1998).

Accepting that these feelings are part of the process is the first step; the second is to begin to deal with them. Selby et al. (2011) suggests that often people don't know how to deal with negative emotions and as a result may exacerbate them. For example, people who are sad tend to withdraw from social activities and the company of others. This tends to isolate them even further and to intensify the emotion.

Research has also shown that individuals who "vent" anger make themselves angrier. So, for example, hitting something or screaming at someone in order to alleviate the anger and "get it out" does not dissipate the anger but instead adds fuel to the fire (Bushman, 2002). Ironically, doing nothing at all when angry has been found to be a more effective way to dissipate the emotion than responding angrily.

What is recommended, instead, is engaging in training to feel and/or think a new way. This means that when the upsetting emotion is experienced, the individual does something to cut down on, distract and eventually replace the negative feeling. For example, individuals who are feeling sad or anxious are encouraged to get active and do something physical like going for a walk or heading to the gym. Another method is to turn one's attention to a hobby which engages and focuses the mind. One veteran of World War II, for example, found that working on something which required deep concentration was one of the few ways he was able to shake the negative feelings, memories, thoughts and emotions which plagued him after returning from combat (Sledge, 2002).

The key is to stick with the training. Doing it once or twice and finding it doesn't help is not a proper indication. Instead, training away negative feelings and thoughts means doing the replacement activity over and over again until it becomes a habit.

j) Readjust expectations

Unmet or unrealistic expectations play a role in burnout. Freudenberger (1989) suggests that many employees, particularly those who have trained for their jobs at educational institutions, are completely unprepared for the realities of the workplace. To combat burnout, he comments, we "need to consider what values our training institutions and universities impart that do not prepare us for what we will find once we enter our professions and work in our chosen fields" (p. 5). Schools tend to emphasize ideals in the job or profession; they rarely prepare their students for all the ways in which they will discover the job has not lived up to their vision of the ideal.

Eager, enthusiastic and idealistic employees, particularly those in the helping professions, tend to have expectations about what they can achieve in the workplace. The problem is, frequently these expectations are more often than not unachievable. The answer is not to dispense with expectations altogether and expect the worst from each day, but rather to make expectations more realistic. As Maslach (2003) suggests,

> . . . you have to work out a list of specific accomplishments that you can shoot for on a given day, a given month – even for the year. . . . The key word here is *specific* – the possible accomplishments have to be well defined in concrete terms. If they refer to specific behaviors, rather than general abstractions, then they are clearly "do-able". . . . Not only should these goals be specific, they should be *realistic* as well. By realistic, I mean that there must be a reasonable chance that you will actually be able to accomplish the goal. If the goal is virtually impossible to achieve, then you are doomed to failure. (pp. 149–50; italic in original)

For Maslach, setting realistic goals allows an employee to understand her limitations as well as her strengths. It also moves an employee away from having expectations or ideals which are likely to lead only to disappointment and disillusionment.

k) Mend fences

One of the unfortunate problems with burnout is that individuals who are burning out generally push others away – including those who want to help. These people include friends, family, and/or co-workers. Pushing them away may have caused damage to important relationships, and when trying to recover from burnout, dealing with those relationships is not always high on the priority list.

Sooner or later, though, some part of recovery must involve mending fences. As Harold Sala (2009) notes, those people alienated along the way may not be terribly excited to see the burned out individual back in their lives. They have learned to live without that individual and things might not go back to the way they were either immediately or ever. Nonetheless, particularly where family is concerned, an effort should be made to mend fences and either to re-establish old relationships and connections or to form new ones.

l) Change the job or find another job

Sometimes a job is not fixable and looking for new work may be in order. This can mean looking for another position within the organization or beyond. It may also mean leaving a particular career or profession and looking for a different type of work entirely.

Most burnout experts, however, urge caution in this regard. Employees who burned out are often not thinking rationally; they are not themselves. They are disillusioned, exhausted and feel they have failed in some way. Quitting without attempting to do everything possible to remedy the situation might well result in an increased sense of failure, and entering a new job without having first recovered, which means starting off on the wrong foot in the new environment.

Instead, burned out employees should first attempt to sort out what the specific problems are in their current job. One question that should be asked is whether the correct problems have been identified or whether there is still a general dissatisfaction that can't be pinpointed? If all the problems have been pinpointed, has any action been taken to try to resolve the sources of stress? Have all possible means of resolving the problem been attempted? Have all the people who might be able to help or alter the job been consulted?

Where all attempts at eliminating the sources of stress or changing the job have failed, however, employees need to think carefully about the next step. What, for example, is the ideal job? Why? Realistically, what are some of the problems that might be associated with the ideal job?

Once again, it's advisable for the employee to seek advice from trusted family, friends and colleagues. It's also advisable to take the time to talk to a career counselor and to individuals who perhaps work in the employee's desired environment. All of this is likely to shed more light on the idea of a new job and to provide enough information in order to determine whether making a move is the solution.

Conclusion: the joy of burnout

Burnout is painful and disruptive, but it also has the potential to lead us to a much happier life. As Glouberman explains,

> [b]urning out may literally save our lives by stopping us before we suffer a more serious or fatal illness. It operates like a circuit breaker that keeps the whole system from blowing. On another level, burnout saves our life by showing us how and when our life lost its old meaning and by forcing us to do something about it. We may not save our old life, but we can free ourselves to be more fully alive. (p. 57)

Coming back from burnout requires that we make changes to our lives, but we know instinctively that we cannot recover without them. Once those changes have been made (whether deliberately or simply because we can no longer continue the way we have been), we discover a joy in living that we thought had been lost in our lives. Burnout has literally saved our lives – it has taken us through the darkest places and deposited us, finally, in a world in which we find more space, more time, more balance and ultimately more perspective.

References

Bernier, D. (1998) A study of coping: Successful recovery from severe burnout and other reactions to severe work-related stress. *Work & Stress*, 12(1), 50–65.

Bruce, M. (2007) Keep a stress diary. *Nursing Standard*, 22(5), 64.

Bushman, B.J. (2002) Does venting anger feed or extinguish the flame? Catharsis, rumination, distraction, anger and aggressive responding. *Personality and Social Psychology Bulletin*, 28, 724–31.

Casserley, T.B., and Megginson, D. (2009) *Learning from Burnout: Developing Sustainable Leaders and Avoiding Career Derailment.* Oxford, UK: Butterworth-Heinemann.

Cherniss, C. (1992) Long term consequences of burnout: An exploratory study. *Journal of Organizational Behaviour*, 13, 1–11.

Clarkson, G.P., and Hodgkinson, G.P. (2007) What can occupational stress diaries achieve that questionnaires can't? *Personnel Review*, 36(5), 684–700.

Freudenberger, H.J. (1989) Burnout: Past, present, and future concerns. In D.T. Wessells, Jr. (Ed.), *Professional Burnout in Medicine and the Helping Profession* (pp. 1–10). New York, New York: Routledge.

Fritz, C., Yankelevich, M., Zarubin, A., and Barger, P. (2010) Happy, healthy, and productive: The role of detachment from work during nonwork time. *Journal of Applied Psychology*, 95(5), 977–83.

Glicken, M.D. (1983) A counseling approach to employee burnout. *Personnel Journal*, 62(3), 222–8.

Glouberman, D. (2002) *The Joy of Burnout.* Isle of Wight: Skyros Books.

Gorter, R.C., Eijkman, M.A.J., and Hoogstraten, J. (2001) A career counseling program for dentists: Effects on burnout. *Patient Education and Counseling*, 43(1), 23–30.

Macdonald, S., Csiernik, R., Durand, P., Rylett, M., and Wild, T.C. (2006) Prevalence and factors related to Canadian workplace health programs. *Canadian Journal of Public Health*, 97(2), 121–5.

Maslach, C. (2003) *Burnout: The Cost of Caring.* Englewood Cliffs, New Jersey: Prentice-Hall. (Original work published 1982.)

Morse, G., Salyers, M., Rollins, A., Monroe-DeVita, M., and Pfahler, C. (2012) Burnout in mental health services: A review of the problem and its remediation. *Administration and Policy in Mental Health and Mental Health Services Research*, 39(5), 341–52.

Pines, A. and Aronson, E. (1988) *Career Burnout: Causes and Cures.* New York, New York: The Free Press.

Psychological counselling recommended for victims of burnout (24 March 1986) *The Ottawa Citizen*, D3.

Randstad Canada (12 July 2012) Work life balance tips to help you define the line between work and home. *Canada NewsWire.*

Sala, H.J. (2009) *Making Your Emotions Work for You: Coping with Stress, Avoiding Burnout, Overcoming Fear . . . and More.* Eugene, Oregon: Harvest House Publishers.

Selby, E.A., Pychyl, T., Marano, H.E., and Jaffe, A. (6 September 2011) Self-Sabotage: The enemy within. *Psychology Today*, 44(5), 55.

Shanafelt, T.D., West, C.P., Sloan, J.A., Novotny, P.J., Poland, G.A., Menaker, R., Rummans, T.A., and Dyrbye, L.N. (2009) Career fit and burnout among academic faculty. *Archives of Internal Medicine*, 169(10), 990–5.

Sledge, E.B. (2002) *China Marine.* Tuscaloosa, Alabama: University of Alabama Press.

Stoltzfus, E.R. (2009) *Access to Wellness and Employee Assistance Programs in the United States.* Bureau of Labor Statistics. Retrieved from *http://www.bls.gov/opub/cwc/cm20090416ar01p1.htm.*

Talbott, S. (2012) *Secret of Vigor: How to Overcome Burnout, Restore Metabolic Balance, and Reclaim Your Natural Energy*. Alameda, California: Hunter House.

Taylor, I. (1 December 2008) Managers play important supporting role in workplace mental health. *Employee Benefit News Canada*, 5(8), 18.

INDEX

Kolibri 360 GmbH
Ferdinand-Jühlke-Straße 7
99095 Erfurt, Deutschland
produktsicherheit@kolibri360.de